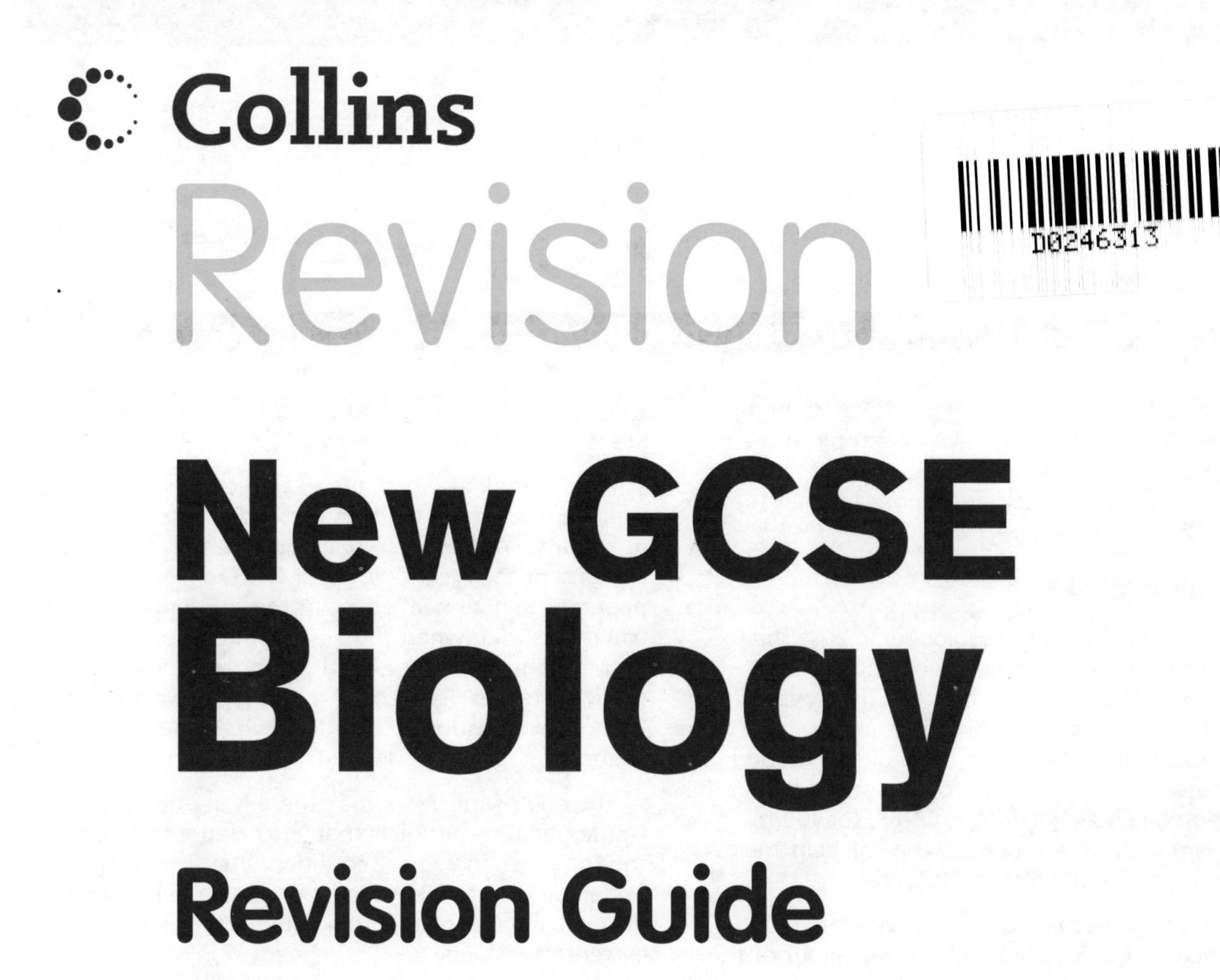

New GCSE Biology

Revision Guide

Higher

For OCR Gateway B

Authors: Colin Bell
Ian Honeysett

About this book

This book covers the content you will need to revise for GCSE Biology for OCR Gateway B at Higher Level. Written by GCSE examiners, it is designed to help you get the best grade in your GCSE Biology exams.

The book is divided into six modules.

Revision and practice

The book is divided into two parts: **Revision guide** and **Workbook**. Begin by revising a topic in the Revision guide section, then test yourself by answering the exam-style questions for that topic in the Workbook section.

Revision guide

The pages in the Revision guide summarise the content of the exam specification and act as a memory jogger. There is a question (**Improve your grade**) on each page that will help you to check your progress. Typical answers to these questions, and examiner's comments, are provided at the end of the Revision guide section (pages 58–63) for you to compare with your responses. This will help you to improve your answers in the future.

At the end of each module, you will find a **Summary** page. This highlights some important facts from each module.

Workbook

The **Workbook** (pages 74–127) allows you to work at your own pace on some typical exam-style questions. You will find that the actual GCSE questions are more likely to test knowledge and understanding across topics. However, the aim of the Revision guide and Workbook is to guide you through each topic so that you can identify your areas of strength and weakness.

The Workbook also contains example questions that require longer answers (**Extended response questions**). You will find one question that is similar to these in each section of your written exam papers. The quality of your written communication will be assessed when you answer these questions in the exam, so practise writing longer answers, using sentences. For ease of use, the **Answers** to all **Workbook** questions are detachable. These can be found on pages 137–144.

At the end of the Workbook there is a series of **Grade booster checklists** that you can use to tick off the topics when you are confident about them and understand certain key ideas. These Grade boosters give you an idea of the grade at which you are currently working.

Additional features

- **Exam tips** give additional exam advice.
- **Remember boxes** pick out key facts to help you revise.
- A **Glossary** allows quick reference to the definitions of the scientific terms highlighted in bold.

Published by Collins
An imprint of HarperCollins*Publishers*
77–85 Fulham Palace Road
Hammersmith
London W6 8JB

Browse the complete Collins catalogue at:
www.collins.co.uk

10 9 8 7 6 5 4 3

ISBN 978-0-00-741611-0

British library Cataloguing in Publication Data

A Catalogue record for this publication is available from the British Library.

Written by Colin Bell and Ian Honeysett.

Project managed by Sally Moon Publishing Services
Design by wired2create
Page make-up by Jordan Publishing Design Limited
Illustrations by Kathy Baxendale, IFA Design Ltd ,
Ken Vail Graphic Design and Nigel Jordan
Edited by Lyn Ward
Printed and bound by Printing Express, Hong Kong

Acknowledgements

Whilst every effort has been made to trace the copyright holders, in cases where this has been unsuccessful, or if any have inadvertently been overlooked, the Publishers will be pleased to make the necessary arrangements at the first opportunity.

Contents

Fitness and health

Blood pressure

- **Blood pressure** is measured in millimetres of mercury. This is written as mmHg.
- Blood pressure has two measurements: **systolic pressure** is the maximum pressure the heart produces and **diastolic pressure** is the blood pressure between heart beats.
- Different factors can cause a person's blood pressure to increase or decrease:
  – It can be increased by stress, high **alcohol** intake, smoking and being overweight.
  – It can be decreased by regular exercise and eating a balanced **diet**.
- High blood pressure can cause blood vessels to burst. This can cause damage to the brain, which is often called a **stroke**. It can also cause damage to the kidneys.
- Low blood pressure can lead to dizziness and fainting as the blood supply to the brain is reduced, and poor circulation to other areas such as the fingers and toes.

Fitness and health

- There is a difference between fitness and health:
  – Fitness is the ability to do physical activity.
  – Health is being free from diseases such as those caused by **bacteria** and **viruses**.
- Your general level of fitness can be measured by your cardiovascular **efficiency**.
- Your fitness can also be measured for different activities:
  – strength, by the amount of weights lifted
  – flexibility, by the amount of joint movement
  – stamina, by the time of sustained exercise
  – agility, by changing direction many times
  – speed, by a sprint race.
- Therefore, you can be very fit for a sprint race but not perform well in a marathon.
- Ways of measuring fitness should be evaluated to check effectiveness in particular situations.

Smoking

- Smoking can increase blood pressure in a number of ways:
  – **Carbon monoxide** in cigarette smoke causes the blood to carry less oxygen. This means that the heart rate increases so that the tissues receive enough oxygen.
  – Nicotine in cigarette smoke directly increases heart rate.

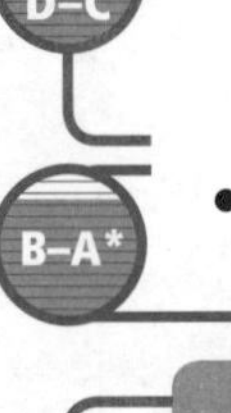

- Carbon monoxide decreases the oxygen-carrying capacity of blood. It combines with **haemoglobin**, preventing it from combining with oxygen, so less oxygen is carried.

Diet and heart disease

- Heart disease is caused by a restricted blood flow to the heart muscle. The risk of getting heart disease is increased by:
  – a high level of saturated fat in the diet, which leads to a build-up of **cholesterol** (a **plaque**) in **arteries**
  – high levels of salt, which can increase blood pressure.
- It is essential to be able to interpret data showing links between the amount of saturated fat eaten, the build-up of cholesterol and the incidence of heart disease.
- The narrowing of the arteries caused by plaques in the coronary arteries can reduce blood flow to heart muscle. The plaques also make blood clots or **thrombosis** more likely to happen, which will also block the artery.

Remember!
It is the blood vessel into the heart muscle that is blocked, not the blood flowing into the heart.

Improve your grade

Heart disease

Excess saturated fat or excess salt in the diet can increase the risk of heart disease. Explain why. *AO1* [4 marks]

Human health and diet

A balanced diet

- It is important for good health to eat a balanced **diet** containing the correct amounts of the chemicals found in food. Three of these are:
  - **carbohydrates**, which are made up of simple sugars such as glucose
  - proteins, which are made up of **amino acids**
  - fats, which are made up of fatty acids and glycerol.
- A balanced diet varies according to factors including age, gender, level of activity, religion, being **vegetarian** or **vegan**, or because of medical issues such as food allergies.

D–C

Chemistry of foods

- If you eat too much fat and carbohydrate, they are stored in the body.
  - Carbohydrates are stored in the liver as glycogen or converted into fats.
  - Fats are stored under the skin and around organs as adipose tissue. Although proteins are essential for growth and repair, they cannot be stored in the body.

B–A*

Protein intake

- Proteins are needed for growth and so it is important to eat the correct amount. This is called the estimated average daily requirement (**EAR**) and can be calculated using the formula:

  EAR in g = 0.6 × body mass in kg
- Sue has a mass of 72.5 kg. Her EAR is 0.6 × 72.5 = 43.5 g/day.
- Too little protein in the diet causes the condition called **kwashiorkor**. This is more common in developing countries due to overpopulation and lack of money to improve agriculture.

D–C

- The EAR is only an estimated figure based on an average person. The EAR for protein might be affected by factors such as body mass, age, pregnancy or breast-feeding (lactation).
- Although proteins cannot be stored in the body, some amino acids can be converted by the body into other amino acids.
- Proteins from meat and fish are called **first-class proteins**. They contain all the essential amino acids that cannot be made by the human body.
- Plant proteins are called **second-class proteins** as they do not contain all the essential amino acids.

B–A*

Overweight or underweight?

- To work out if a person is overweight or underweight, calculate their **body mass index** (BMI):

$$\text{BMI} = \frac{\text{mass in kg}}{(\text{height in m})^2}$$

Tom is 170 cm tall and has a mass of 80 kg. Calculate his BMI.

170 cm = 1.7 m

$$\text{BMI} = \frac{80}{1.7^2} = 27.7$$

EXAM TIP

The height must be in metres and the mass in kilograms.

D–C

- A BMI of more than 30 means the person is **obese**, 25–30 is overweight, 20–25 is normal, less than 20 is underweight. With a BMI of 27.7, Tom is overweight.
- Some people may become ill as they choose to eat less than they need. This may be caused by low self-esteem, poor self-image or a desire for what they think is perfection.

Improve your grade

EAR for protein

Explain the importance of knowing your EAR for protein. *AO1/2* [4 marks]

Staying healthy

Malaria

D–C

- Malaria is caused by a protozoan called *Plasmodium*, which feeds on human **red blood cells**.
- *Plasmodium* is carried by mosquitoes, which are **vectors** (i.e. not affected by the disease), and transmitted to humans by mosquito bites.
- *Plasmodium* is a **parasite** and humans are its host. A parasite is an organism that feeds on another living organism, causing it harm.

B–A*

- Knowledge of the mosquito's life cycle has helped to stop the spread of malaria (by draining stagnant water, putting oil on the water surface and spraying **insecticide**). This knowledge has also helped to develop new treatments for malaria.

Cancers

D–C

- Changes in lifestyle and **diet** can reduce the risk of some **cancers**:
  - Not smoking reduces the risk of lung cancer.
  - Using sunscreen reduces the risk of skin cancer.

B–A*

- Benign **tumour** cells, such as in warts, divide slowly and are harmless.
  Cancers are malignant tumours: the cells display uncontrolled growth and may spread.
- Ways of interpreting data on cancer and survival/mortality rates should be considered.

Remember!
Different antibodies are required to deal with different pathogens.

The fight against illness

D–C

- **Pathogens** (disease-causing organisms) produce the symptoms of an infectious disease by damaging the body's cells or producing poisonous waste products called **toxins**.
- The body protects itself by producing **antibodies**, which lock onto **antigens** on the surface of pathogens such as a **bacterium**. This kills the pathogen.
- Human **white blood cells** produce antibodies, resulting in **active immunity**. This can be a slow process but has a long-lasting effect. Vaccinations using antibodies from another human or animal result in passive immunity, which has a quick but short-term effect.

B–A*

- Each pathogen has its own antigens, so a specific antibody is needed for each pathogen.
- The process of immunisation is also called vaccination:
  - It starts with injecting a harmless pathogen carrying antigens.
  - The antigens trigger a response by white blood cells, producing the correct antibodies.
  - Memory cells (a type of T-lymphocyte cell) remain in the body, providing long-lasting immunity to that disease.
- Immunisation carries a small risk to the individual, but it avoids the potentially lethal effect of the pathogen, as well as decreasing the risk of spreading the disease.

Treatments and trials

D–C

- **Antibiotics** (against bacteria and **fungi**) and **antiviral drugs** (against **viruses**) are specific in their action.
- An antibiotic destroys a pathogen; an antiviral drug slows down the pathogen's development.
- New treatments, such as vaccinations, are tested using animals, human tissue and computer models before human **trials**. Some people object to causing suffering in animals in such tests.

B–A*

- A **placebo** is a harmless pill. Placebos are used as a comparison in drug testing so the effect of a new drug can be assessed.
- In a **blind trial**, the patient does not know whether they are receiving a new drug or a placebo. In a double-blind trial, neither the patient nor the doctor know which treatment is being used. These types of trials avoid a 'feel-good factor' and a biased opinion.
- Excessive use of antibiotics has resulted in resistant forms of bacteria being more common than non-resistant forms. For example, resistant MRSA has thrived, causing serious illness.

Improve your grade

Immunisation

Immunisation carries a small risk. Despite this risk, why is immunisation important? *AO1* [4 marks]

The nervous system

How do eyes work

D–C

- The main parts of the eye have special functions.
- Light rays are **refracted** (bent) by the cornea and lens.
- The retina contains light receptors. Some are sensitive to different colours.
- **Binocular vision** helps to judge distance by comparing the images from each eye; the more different they are, the nearer the object.

B–A*

- The eye can focus light from distant or near objects by altering the shape of the lens. This is called **accommodation**.
- To focus on distant objects, the ciliary muscles relax, and the suspensory ligaments tighten, so the lens has a less rounded shape.
- To focus on near objects, the ciliary muscles contract and the suspensory ligaments slacken, so the lens regains a more rounded shape due to its elasticity.

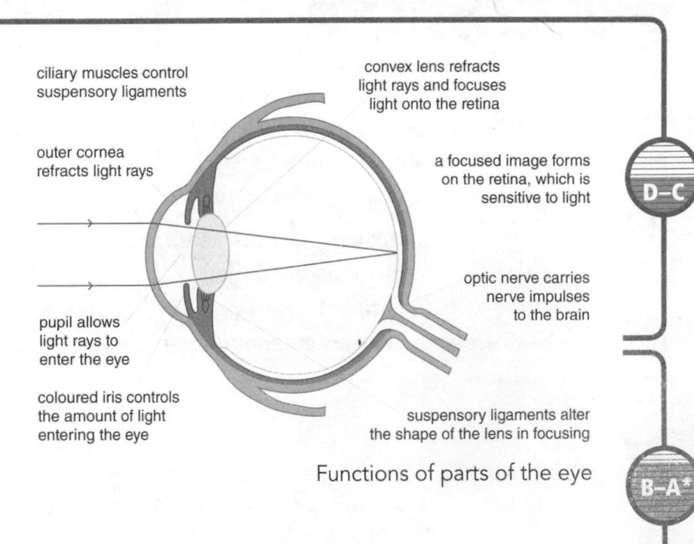

Functions of parts of the eye

Faults in vision

D–C

- Red-green colour blindness is caused by a lack of specialised cells in the retina.
- Long and short sight are caused by the eyeball or lens being the wrong shape. In long sight, the eyeball is too short or the lens is too thin, so the image is focused behind the retina. In short sight, the eyeball is too long or the lens is too rounded so the lens refracts light too much, so the image would be focused in front of the retina.

B–A*

- Corneal surgery or a lens in glasses or contact lenses corrects long and short sight. A convex lens is used to correct long sight, a concave lens to correct short sight.

Nerve cells

D–C

- Nerve cells are called neurones. Nerve impulses pass along the **axon**.
- What happens in a **reflex** action is shown by a reflex arc. The links in a reflex arc are:

  stimulus→receptor→**sensory neurone**→**central nervous system**→**motor neurone**→effector→response
- The pathway for a spinal reflex is:

  receptor→sensory neurone→relay neurone→motor neurone→effector

B–A*

- Neurones are adapted by being long, having branched endings (dendrites) to pick up impulses and having an insulator sheath.
- The gap between neurones is called a **synapse**. The arrival of an impulse triggers the release of a transmitter substance, which **diffuses** across the synapse. The transmitter substance binds with receptor **molecules** in the membrane of the next neurone causing the impulse to continue.

branching dendrites
muscle fibres (effector)
cell body
axon
nucleus
sheath
a motor neurone

Parts of a motor neurone

Improve your grade

Vision

In old age, muscles lose their ability to contract and relax quickly. Ligaments become less flexible. The lens becomes less elastic. Explain the effects of these changes on vision. *AO2* [4 marks]

Remember!
In the eye the light rays are refracted, *not* reflected.

Drugs and you

Types of drugs

D–C

- Drugs have a legal classification. Class A drugs are the most dangerous and have the heaviest penalties. Class C drugs are the least dangerous, with the lightest penalties.
- There are different types of drugs:
  - **depressants** (**alcohol**, solvents, temazepam)
  - **painkillers** (aspirin, paracetamol)
  - **stimulants** (nicotine, MDMA ('ecstasy'), caffeine)
  - **performance enhancers** (anabolic steroids)
  - **hallucinogens** (LSD).

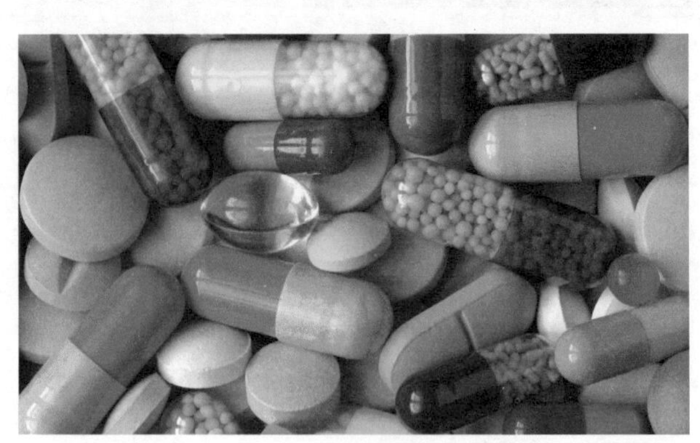

Different prescription drugs

B–A*

- Depressants block the transmission of nerve impulses across **synapses** by binding with receptor molecules in the membrane of the receiving neurone.
- Stimulants cause more neurotransmitter substances to cross synapses.

Effects of smoking

D–C

- Cigarette smoke contains many chemicals that stop cilia moving.
- Cilia (tiny hairs) are found in the epithelial lining of the trachea, bronchi and bronchioles.
- A 'smoker's cough' is a result of
  - dust and particulates in cigarette smoke collecting and irritating the epithelial lining
  - mucus not being moved by the cilia.

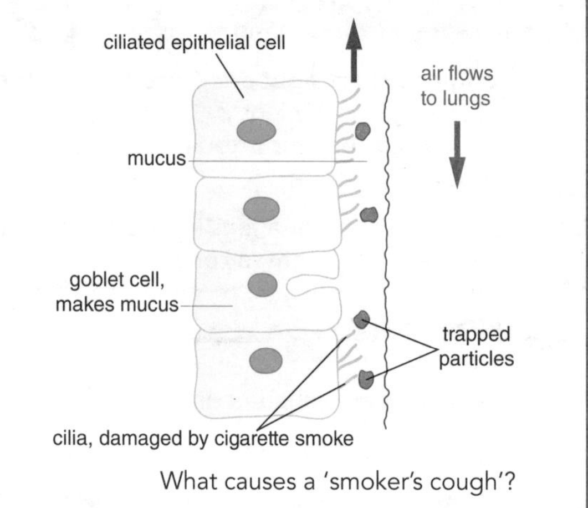

What causes a 'smoker's cough'?

EXAM TIP

Avoid the common error of writing 'the rate of lung cancer is affected' without explaining that it increases or decreases.

Effects of alcohol

D–C

- The alcohol content of alcoholic drinks is measured in **units of alcohol**.
- Drinking alcohol increases reaction times and increases the risk of accidents.

B–A*

- The liver is damaged when it breaks down **toxic** chemicals such as alcohol. This is called cirrhosis of the liver.

How science works

You should be able to:

- identify complex relationships between variables such as in the interpretation of data related to alcohol content, accident statistics and reaction times.

Improve your grade

Affects of alcohol

Drinking alcohol increases the risk of accidents. Explain why. *AO2* [4 marks]

Staying in balance

Homeostasis

- Keeping a constant internal environment is called homeostasis.
- Homeostasis involves balancing bodily inputs and outputs.
- Automatic control systems keep the levels of **temperature**, water and **carbon dioxide** steady. This makes sure all cells can work at their optimum level.

D–C

- Negative feedback controls are used in homeostasis. Negative feedback systems act to cancel out a change such as a decreasing temperature level.

B–A*

Remember!
Heat stroke, dehydration and hypothermia can be fatal.

Temperature control

- The body temperature of 37 °C is linked to the **optimum temperature** for many **enzymes**.
- A high temperature can cause:
  - **heat stroke** (skin becomes cold and clammy and pulse is rapid and weak)
  - **dehydration** (loss of too much water).
- Both heat stroke and dehydration can be fatal if not treated.
- To avoid overheating, sweating increases heat transfer from the body to the environment.
- The **evaporation** of sweat requires body heat to change the liquid sweat into water vapour.
- A very low temperature can cause **hypothermia** (slow pulse rate, violent shivering), which can be fatal if not treated.

D–C

- Blood temperature is monitored by the **hypothalamus** gland in the brain. Reaction to temperature extremes are controlled by the nervous and hormonal systems, which trigger vasoconstriction or vasodilation.
- Vasoconstriction is the constriction (narrowing) of small blood vessels in the skin. This causes less blood flow and less heat transfer.
- Vasodilation is the dilation (widening) of small blood vessels in the skin. This causes more blood flow near the skin surface resulting in more heat transfer.

B–A*

Vasoconstriction. When the body is too cold small blood vessels in the skin constrict and so less blood flows through them, reducing heat loss.

Vasodilation. When the body is too hot small blood vessels in the skin dilate and so blood flow increases, bringing more blood to the surface, where it loses heat.

sweat evaporates from the skin surface, cooling it

Vasoconstriction and vasodilation

EXAM TIP

In vasodilation and vasoconstriction, the *size* of the small blood vessels is altered. The blood vessels do *not* move.

Control of blood sugar levels

- A **hormone** called **insulin** controls **blood sugar levels**.
- Hormone action is slower than nervous reactions as the hormones travel in the blood.
- Type 1 diabetes is caused by the pancreas not producing any insulin, so must be treated by doses of insulin. Type 2 diabetes, which is caused either by the body producing too little insulin or the body not reacting to it, can be controlled by **diet**.

D–C

- Insulin converts excess glucose in the blood into glycogen, which is stored in the liver. This regulates the blood sugar level.
- The insulin dosage in Type 1 diabetes needs to vary according to the person's diet and activity. Strenuous exercise needs more glucose to be present in the blood, so a lower insulin dose is required.

B–A*

Improve your grade

Types of diabetes

Joe has Type 1 diabetes and Charlie has Type 2 diabetes. Explain why they require different treatments. *AO1* [4 marks]

Controlling plant growth

Plant responses

- **Phototropism** is a plant's growth response to light. **Geotropism** is a plant's growth response to **gravity**.
- Parts of a plant respond in different ways:
  - Shoots are positively phototropic (they grow towards light) and negatively geotropic (they grow away from the pull of gravity).
  - Roots are negatively phototropic (they grow away from light) and positively geotropic (they grow with the pull of gravity).

Remember!
A positive reaction means that the root or shoot grows towards the stimulus.

Plant hormones (auxins)

- **Auxins** are a group of **plant hormones**. They move through the plant in solution.
- Auxins are involved in phototropism and geotropism.
- Auxins are made in the root and shoot tip.
- Different amounts of auxin are found in different parts of the shoot when the tip is exposed to light. More auxin is found in the shady part of shoots. A higher amount of auxin will increase the length of cells. Therefore the increase in cell length on the shady side of the shoot causes curvature of the shoot towards the light.

direction of light

When the shoot tip is removed the shoot does not grow

Commercial uses of plant hormones

- Plant hormones have many commercial uses. They are used:
  - as selective weedkillers, which kill specific weeds and increase crop yield
  - as rooting powder to increase root growth of cuttings
  - to delay or accelerate fruit ripening to meet market demands
  - to control dormancy in seeds.

Spraying crops with selective weedkiller

How science works

You should be able to:

- interpret data from phototropism experiments on auxin action.

Improve your grade

Phototropism

Plant shoots grow straight upwards in the dark but will grow towards a light source. Explain how and why they do this. *AO1* [4 marks]

Variation and inheritance

Inherited characteristics

D–C

- Some human characteristics, such as facial features, can be inherited. They can be **dominant** or **recessive.**
- **Alleles** are different versions of the same **gene**.

B–A*

- There is debate over how much 'nature or nurture' (genetic or environmental factors) affect intelligence, sporting ability and health.
- Dominant and recessive characteristics depend on dominant and **recessive alleles**.
- Dominant alleles are expressed when present but recessive alleles are expressed only in the absence of the dominant allele.

Remember!
A gene can have two different alleles, one dominant and the other recessive.

Chromosomes

D–C

- Most body cells have the same number of **chromosomes**. The number depends on the **species** of organism. Human cells have 23 pairs.
- **Sex chromosomes** determine sex in **mammals**. Females have identical sex chromosomes called XX, males have different sex chromosomes called XY.
- A sperm will carry either an X or a Y chromosome. All eggs will carry an X chromosome.

B–A*

- There is a **random** chance of which sperm **fertilises** an egg. There is therefore an equal chance of the offspring being male or female.

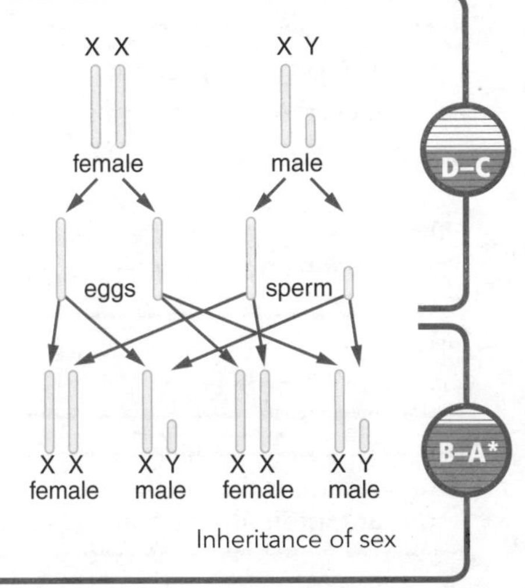

Inheritance of sex

Genetic variation

D–C

- Genetic **variation** is caused by:
  - **mutations**, which are random changes in genes or chromosomes
  - rearrangement of genes during the formation of **gametes**
  - **fertilisation**, which results in a zygote with alleles from the father and mother.

A monohybrid cross

B–A*

- A monohybrid cross involves only one pair of characteristics controlled by a single gene, one allele being dominant and one recessive.
- **Homozygous** means having identical alleles, **heterozygous** means having different alleles.
- A person's genotype is their genetic make-up. Their phenotype is which alleles are expressed.

Inherited disorders

D–C

- Inherited disorders are caused by faulty genes.
- Many personal and ethical issues are raised:
  - in deciding to have a genetic test (a positive result could alter lifestyle, career, insurance)
  - by knowing the risks of passing on an inherited disorder (whether to marry/have a family).

B–A*

- Inherited disorders are caused by faulty alleles, most of which are recessive.
- It is possible to predict the probability of inheriting such disorders by interpreting genetic diagrams.

Improve your grade

Inherited disorders

Rabeena finds out that her mother has cystic fibrosis, an inherited disorder caused by a faulty allele (c). Her father does not have this faulty allele. Explain why Rabeena hopes her future husband will not be heterozygous for cystic fibrosis. *AO2/3* [4 marks]

B1 Summary

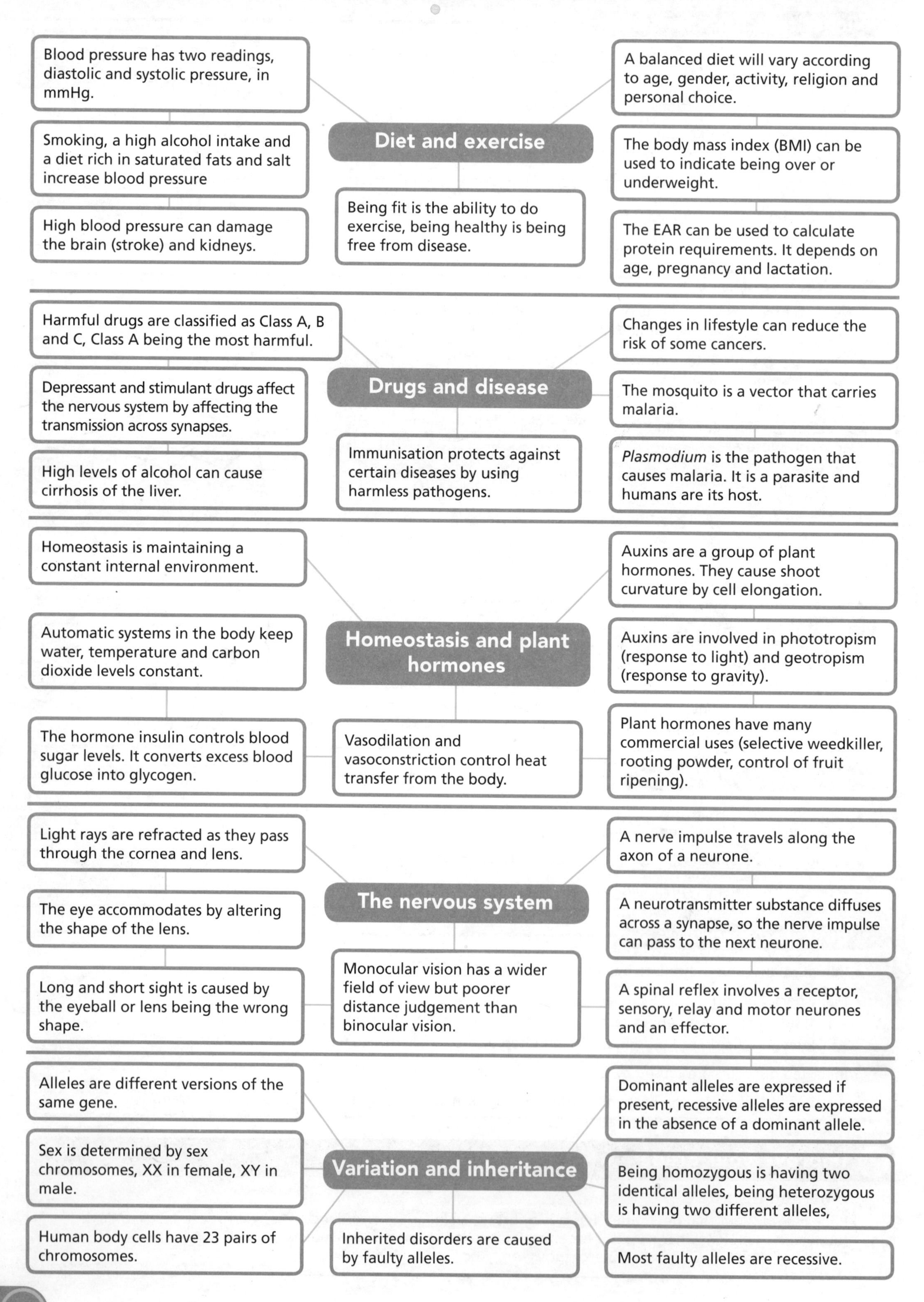

Diet and exercise
Blood pressure has two readings, diastolic and systolic pressure, in mmHg.
Smoking, a high alcohol intake and a diet rich in saturated fats and salt increase blood pressure
High blood pressure can damage the brain (stroke) and kidneys.
Being fit is the ability to do exercise, being healthy is being free from disease.
A balanced diet will vary according to age, gender, activity, religion and personal choice.
The body mass index (BMI) can be used to indicate being over or underweight.
The EAR can be used to calculate protein requirements. It depends on age, pregnancy and lactation.
Drugs and disease
Harmful drugs are classified as Class A, B and C, Class A being the most harmful.
Depressant and stimulant drugs affect the nervous system by affecting the transmission across synapses.
High levels of alcohol can cause cirrhosis of the liver.
Immunisation protects against certain diseases by using harmless pathogens.
Changes in lifestyle can reduce the risk of some cancers.
The mosquito is a vector that carries malaria.
Plasmodium is the pathogen that causes malaria. It is a parasite and humans are its host.
Homeostasis and plant hormones
Homeostasis is maintaining a constant internal environment.
Automatic systems in the body keep water, temperature and carbon dioxide levels constant.
The hormone insulin controls blood sugar levels. It converts excess blood glucose into glycogen.
Vasodilation and vasoconstriction control heat transfer from the body.
Auxins are a group of plant hormones. They cause shoot curvature by cell elongation.
Auxins are involved in phototropism (response to light) and geotropism (response to gravity).
Plant hormones have many commercial uses (selective weedkiller, rooting powder, control of fruit ripening).
The nervous system
Light rays are refracted as they pass through the cornea and lens.
The eye accommodates by altering the shape of the lens.
Long and short sight is caused by the eyeball or lens being the wrong shape.
Monocular vision has a wider field of view but poorer distance judgement than binocular vision.
A nerve impulse travels along the axon of a neurone.
A neurotransmitter substance diffuses across a synapse, so the nerve impulse can pass to the next neurone.
A spinal reflex involves a receptor, sensory, relay and motor neurones and an effector.
Variation and inheritance
Alleles are different versions of the same gene.
Sex is determined by sex chromosomes, XX in female, XY in male.
Human body cells have 23 pairs of chromosomes.
Inherited disorders are caused by faulty alleles.
Dominant alleles are expressed if present, recessive alleles are expressed in the absence of a dominant allele.
Being homozygous is having two identical alleles, being heterozygous is having two different alleles,
Most faulty alleles are recessive.

Classification

Grouping organisms

- All organisms are classified into a number of different groups, starting with their kingdom and finishing with their **species**.
- The groups are: kingdom, phylum, class, order, family, genus and species.
- As you move down towards 'species', there are fewer organisms within each group and they share more similarities.

Remember!
Try remembering the order of the groups by using the first letter of each one.

D–C

- Organisms can be classified in two ways:
  – An *artificial system* is based on one or two characteristics that make identification easier, for example birds that always live by or on the sea can be called seabirds.
  – A *natural system* is based on **evolutionary** relationships and is much more detailed. Animals that are more closely related are more likely to be in the same group.
- Sequencing the bases in **DNA** has enabled scientists to know much more about how closely related organisms are, and has often meant that organisms can be reclassified.

B–A*

Species

- A species is a group of organisms that can interbreed to produce fertile offspring.
- All organisms are named by the **binomial system**. The system works like this:
  – There are two parts to the name, the first is the genus and the second the species.
  – The genus part starts with a capital letter; the species part starts with a lower-case letter.

D–C

Problems with classifying

- Living things are at different stages of evolution, and new ones are being discovered all the time. This makes it difficult to place organisms into distinct groups. An example of this is *Archaeopteryx*. This creature had characteristics that would put it into two different groups:
  – It had feathers, like a bird.
  – It also had teeth and a long, bony tail, like a **reptile**.

D–C

- Some organisms present specific problems:
  – **Bacteria** do not interbreed, they reproduce **asexually**, so they cannot be classified into different species using the 'fertile offspring' idea.
  – Mules are **hybrids**, produced when members of two species (a donkey and a horse) interbreed. Hybrids are infertile, so mules cannot be classed as a species.

B–A*

Classification and evolution

- Organisms that are grouped together are usually closely related and share a recent common ancestor. However, they may have different features if they live in different **habitats**.

D–C

- When classifying organisms, it is important to bear in mind that similarities and differences between organisms may have different explanations:
  – Dolphins have similarities to fish because they live in the same habitat (ecologically related). However, they are classified differently – dolphins are **mammals**.
  – Dolphins and bats have evolved to live in different habitats, but both are mammals – they are related through evolution.

B–A*

How science works

You should be able to:

- explain how a scientific idea has changed as new evidence has been found.

Improve your grade

Classifying newly discovered organisms

Two similar types of animals have been discovered living close together in a jungle.
Describe how scientists could find out how closely related the two animals are.
AO2 [3 marks]

Energy flow

Pyramids of biomass

- Pyramids of numbers and pyramids of **biomass** can both be used to represent feeding relationships between organisms in a food chain or web.

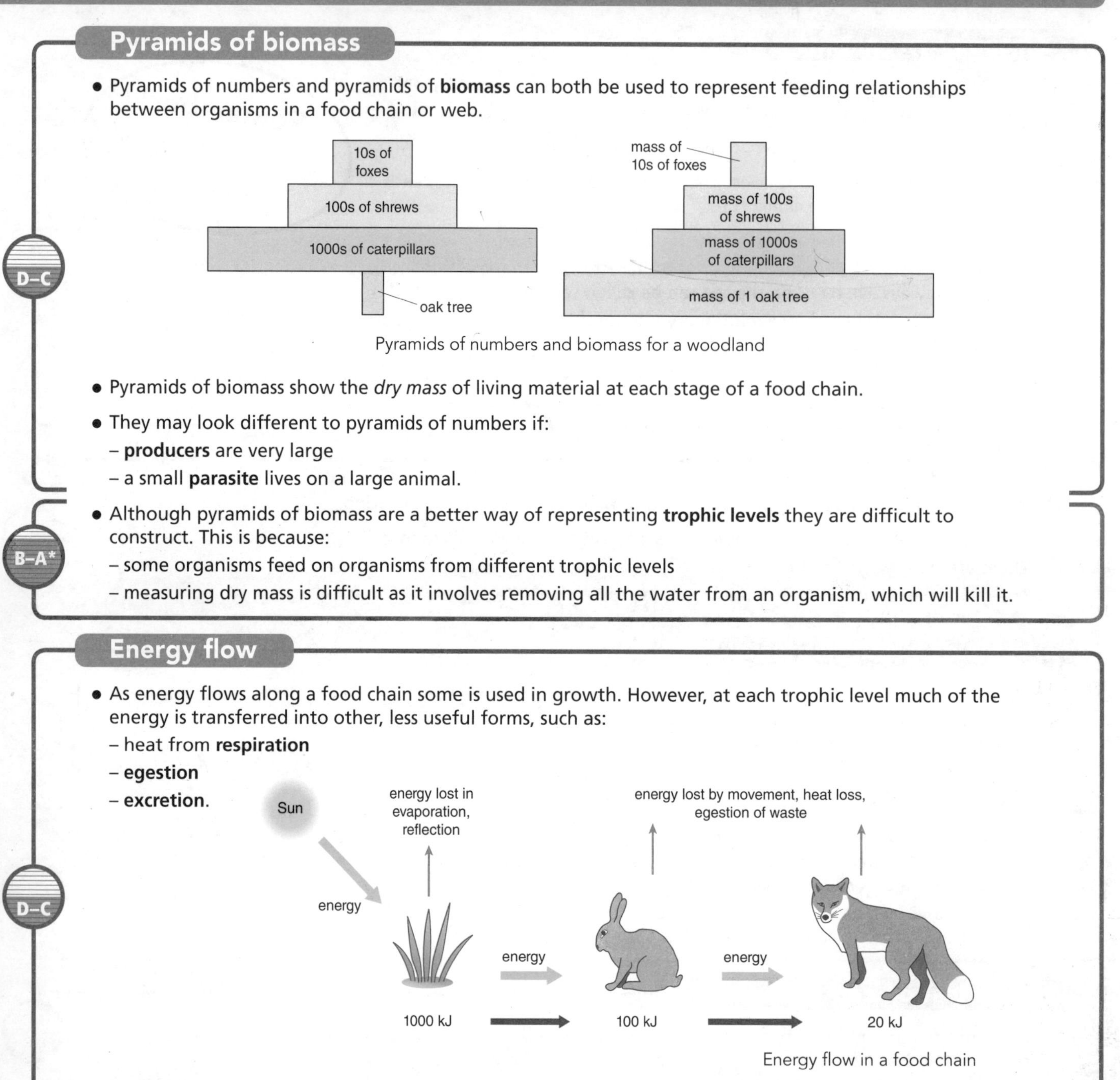

Pyramids of numbers and biomass for a woodland

- Pyramids of biomass show the *dry mass* of living material at each stage of a food chain.
- They may look different to pyramids of numbers if:
  - **producers** are very large
  - a small **parasite** lives on a large animal.
- Although pyramids of biomass are a better way of representing **trophic levels** they are difficult to construct. This is because:
  - some organisms feed on organisms from different trophic levels
  - measuring dry mass is difficult as it involves removing all the water from an organism, which will kill it.

Energy flow

- As energy flows along a food chain some is used in growth. However, at each trophic level much of the energy is transferred into other, less useful forms, such as:
  - heat from **respiration**
  - **egestion**
  - **excretion**.

Energy flow in a food chain

- The material that is lost at each stage of the food chain is not wasted. Most of the waste is used by **decomposers** that can then start another food chain.
- Because each trophic level 'loses' up to 90 per cent of the available energy, an animal at the end of a long food chain does not have much food available to it.
- The **efficiency** of energy transfer can be calculated between trophic levels:

What is the efficiency of energy transfer between the rabbit and the fox?

$$\frac{\text{energy used for growth}}{\text{energy input}} = \frac{20}{100} = 0.2 \text{ or } 20\%$$

B–A*

Improve your grade

Pyramids of biomass

Explain one advantage and one disadvantage of using pyramids of biomass to show feeding relationships. *AO2* [3 marks]

Recycling

The carbon cycle

- **Carbon** is one of a number of elements that are found in living organisms.
- Carbon needs to be **recycled** so it can become available again to other living organisms.
- **Carbon dioxide** is removed from the air by **photosynthesis** in plants.
- Feeding passes carbon compounds along a food chain or web.
- Carbon dioxide is released into the air by:
  - plants and animals **respiring**
  - soil bacteria and **fungi** acting as **decomposers**
  - the burning of **fossil fuels** (**combustion**).

Remember!
Carbon dioxide can be locked up in limestone for a long time. The oceans are often called a carbon sink.

D–C

- Carbon dioxide is also **absorbed** from the air by oceans. Marine organisms make shells made of carbonate, which become **limestone** rocks.
- The carbon in limestone can return to the air as carbon dioxide during volcanic eruptions or weathering.

B–A*

The nitrogen cycle

- Plants take in nitrogen as nitrates from the soil to make protein for growth.
- Feeding passes nitrogen compounds along a food chain or web.
- The nitrogen compounds in dead plants and animals are broken down by decomposers and returned to the soil.

D–C

- A number of **microorganisms** are responsible for the recycling of nitrogen:
  - Decomposers are soil **bacteria** and fungi and they convert proteins and urea into ammonia.
  - **Nitrifying bacteria** convert the ammonia to nitrates.
  - **Denitrifying bacteria** convert nitrates to nitrogen gas.
  - **Nitrogen-fixing bacteria** living in root nodules (or in the soil) fix nitrogen gas – this also occurs by the action of lightning.

B–A*

EXAM TIP
Diagrams of the carbon or nitrogen cycle may not look exactly like this one. Just look for where the arrows go to and from.

The nitrogen cycle

Keeping decomposers working

- For decomposers to break down dead material in soil, they need oxygen and a suitable **pH**.
  - **Decay** will therefore be slower in waterlogged soils as there will be less oxygen.
  - Acidic conditions will also slow down decay.

D–C

Improve your grade

The nitrogen cycle

Explain how nitrogen in a protein molecule in a dead leaf can become available again to a plant. *AO1* [4 marks]

Interdependence

Competition

D–C

- Similar animals living in the same **habitat** compete with each other for resources (e.g. food).
- If they are members of the same **species** they will also compete with each other for mates so they can breed.

B–A*

- An **ecological niche** describes the habitat that an organism lives in and also its role in the habitat. For example, ladybirds live on trees such as sycamore and eat greenfly.
- Organisms that share similar niches are more likely to compete, as they require similar resources. The harlequin ladybird arrived in Britain in 2004 and competes strongly with native ladybirds.
- Competition can be *interspecific* and *intraspecific*:
  - Interspecific is between organisms of different species.
  - Intraspecific is between organisms of the same species and is likely to be more significant as the organisms share more similarities and so need the same resources.

Predator–prey relationships

D–C

- Both **predator** and **prey** show cyclical changes (ups and downs) in their numbers. This is because:
  - When there are lots of prey, more predators survive and so their numbers increase.
  - This means that the increased number of predators eat more prey, so prey numbers drop.
  - More predators starve and so their numbers drop.

B–A*

- The predator peaks occur soon after the peaks of the prey. This is because it takes a little while for the increased supply of food to allow more predators to survive and reproduce.

Parasitism and mutualism

D–C

- As well as competing with each other or eating each other, organisms of different species can also be dependent on each other in other ways.
- **Parasites** feed on or in another living organism called the host.
  - The host suffers as a result of the relationship.
  - Fleas are parasites living on a host (which may be human).
  - Tapeworms are also parasites feeding in the digestive systems of various animals.
- Sometimes both organisms benefit as a result of their relationship. This is called **mutualism**.
  - Insects visit flowers and so transfer pollen, allowing **pollination** to happen. They are 'rewarded' by sugary nectar from the flower.
  - On some coral reefs 'cleaner' fish are regularly visited by larger fish. The large fish benefit by having their parasites removed by the cleaner fish and the cleaner fish gain food.

Remember!
A successfully adapted parasite does not kill its host quickly, as it would then need to find another one.

B–A*

- Pea plants and certain types of bacteria also benefit from mutualism. Pea plants are **legumes** with structures on the roots called nodules. In these nodules live **nitrogen-fixing bacteria**.
  - The bacteria turn nitrogen into nitrogen-containing chemicals and give some to the pea.
  - The pea plant gives the bacteria some sugars that have been produced by **photosynthesis**.

How science works

You should be able to:

- Identify some arguments for and against a scientific or technological development in an unfamiliar situation, in terms of its impact on different groups of people or the environment.

Improve your grade

Niches

The scientist Gauss put forward a theory that said organisms of two different species cannot share an identical ecological niche.

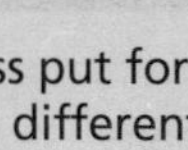

(a) Explain what is meant by the term 'ecological niche'. *AO1* [2 marks]

(b) Suggest why Gauss said that two species cannot share the same niche. *AO2* [2 marks]

Adaptations

Adapting to the cold

D–C

- Some animals are adapted to living in very cold conditions. They keep warm by reducing heat loss. Some have anatomical **adaptations** to help reduce heat loss.
- So:
  - They have excellent **insulation** to cut down heat loss. The arctic fox has thick fur that traps plenty of air for insulation. Seals have thin fur but a thick layer of fat under the skin.
  - These animals are usually quite large, with small ears. This helps to decrease heat loss by decreasing the surface area to volume ratio.
- Animals may try to avoid the cold by changing their behaviour. Some migrate long distances to warmer areas. Others slow down all their body processes and hibernate.

B–A*

- Penguins have a counter-current heat exchange mechanism to help reduce heat loss. The warm blood entering the flippers warms up the cold blood leaving, to stop it cooling the body.
- Other organisms that live in cold climates may have biochemical adaptations, such as antifreeze proteins in their cells.

Adapting to hot, dry conditions

D–C

- Organisms such as camels and cacti live in deserts, in very hot, dry conditions.
- To increase heat loss, animals adapt in a variety of ways.
  - Some are anatomical adaptations, for example camels increase the loss of heat by having very little hair on the underside of their bodies. Animals that live in hot areas are usually smaller and have larger ears than similar animals that live in cold areas. These factors give them a larger surface area to volume ratio, so that they can lose more heat.
  - Other adaptations to lose more heat are behavioural, such as panting or licking their fur.
- To reduce heat gain, animals may change their behaviour, for example they seek shade during the hotter hours around the middle of the day.
- To cope with dry conditions, organisms have behavioural, anatomical and physiological adaptations. For example:
  - Camels can survive with little water because they can produce very concentrated urine.
  - Cacti reduce water loss because their leaves have been reduced to spines. They also have deep roots and can store water in the stem.

B–A*

- Organisms that can survive in hot conditions are called extremophiles. Some bacteria can live in hot springs as they have **enzymes** that do not denature at **temperatures** as high as 100°C.

Spines on a cactus

Specialists or generalists

B–A*

- Some organisms, like polar bears, are called *specialists*, as they are very well adapted to living in specific **habitats**. They would struggle to live elsewhere.
- Others, for example rats, can live in several habitats.
  - These organisms are called *generalists*.
  - They will lose to the specialists in certain habitats.

EXAM TIP

You may have to write about different organisms. Just apply the same principles.

Improve your grade

Living in hot, dry conditions

An elephant has a large body, large ears, skin with few hairs and the ability to produce concentrated urine. Explain which of these features are advantages or disadvantages when living in hot, dry areas. *AO2* [4 marks]

Natural selection

Charles Darwin and natural selection

D–C

- Over 150 years ago Charles Darwin wrote his theory of **natural selection** to explain how **evolution** might happen. It says that if animals and plants are better adapted to their environment, they and the following generations are more likely to survive.
- He did not know exactly how **adaptations** were passed on. We now know that when organisms reproduce, their **genes** are passed on to the next generation.
- The modern version of natural selection can be summarised like this:
  - Within any **species** there is **variation**.
  - Organisms produce far more young than will survive, so there is competition for limited resources such as food.
  - Only those best adapted will survive, which is called *survival of the fittest*.
  - Those that survive pass on successful adaptations to the next generation in their genes.

B–A*

- Over time, the changes produced by natural selection may result in a new species.
  - This only happens if different groups of organisms cannot mate for a long time.
  - The organisms might be prevented from mating because they live in different areas. This is geographical isolation. They might be prevented from mating because of behavioural isolation.
  - If each group evolves differently they might over time become different enough to be classified as separate species.

Modern examples of natural selection

D–C

- Natural selection is difficult to study because it usually takes thousands of years to see the effect. Some examples have been studied over shorter time spans:
  - More and more **bacteria** are developing resistance to **antibiotics**.
  - Peppered moths are dark or pale in colour. Dark moths are better camouflaged in polluted areas, so more of them survive.

Light and dark peppered moths on a tree trunk

Arguments over natural selection

D–C

- At first many people disagreed with Darwin's ideas:
  - Some people thought he did not have enough evidence to back up his theory.
  - Many people disagreed because they thought God had created all species.
- Now Darwin's theory is much more widely accepted. This is because:
  - it explains lots of observations
  - it has been discussed and tested by a wide range of scientists.

B–A*

- There have been other attempts to explain evolution. Before Darwin, Jean Baptiste de Lamarck had a different theory, called the law of acquired characteristics. This said, for example, that giraffes acquired long necks to feed, and this characteristic was passed on.
- As we have discovered more about how genes are passed on, theories like Lamarck's have been proved incorrect and Darwin's theory has become more widely accepted.

EXAM TIP

You need to be prepared to explain the evolution of any organism using Darwin's main ideas.

Improve your grade

Explanations for evolution

Human ancestors had more hair than modern humans. This could be explained by saying that scratching the skin due to parasites has gradually over many generations made some of the hair fall out.

(a) Explain why this theory uses Lamarck's ideas. *AO2* [2 marks]

(b) How might Darwin's ideas be used to explain why modern humans have less hair? *AO2* [2 marks]

Population and pollution

Pollution

- There are many different types of **pollution**. Three that have caused much concern are:
  - **carbon dioxide**, from increased burning of **fossil fuels**, which may increase the greenhouse effect and **global warming**
  - **CFCs**, from aerosols, which destroy the **ozone layer**
  - sulfur dioxide, from burning fossil fuels, which causes **acid rain**.

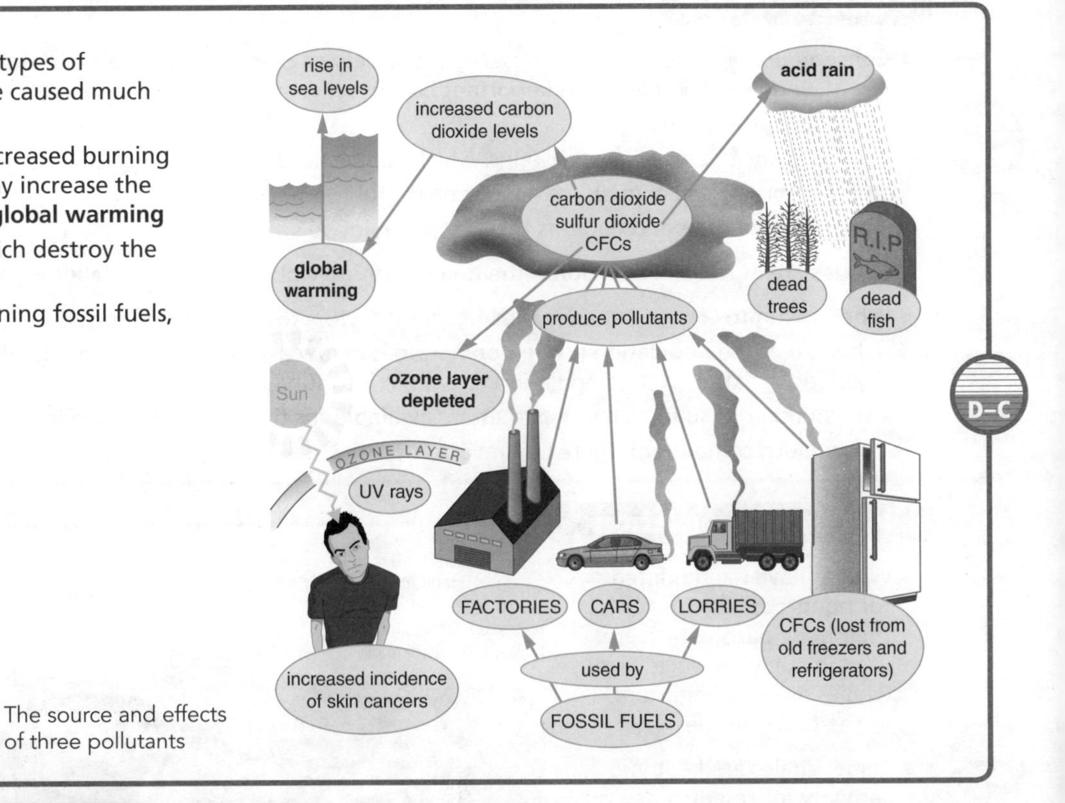

The source and effects of three pollutants

D–C

Population and pollution

- The human **population** of the world is growing at an ever-increasing rate. This is called **exponential growth**.
- This growth in population is happening because the birth rate is exceeding the death rate.

D–C

- The greatest rise in world population figures is occurring in under-developed land masses, such as Africa and India. However, the developed world uses the most fossil fuels (principally the USA and Europe).
- The amount of pollution caused per person or organisation is called the '**carbon footprint**.' This measures the total **greenhouse gas** given off by a person or organisation within a certain time.

B–A*

Measuring pollution

- Pollution in water or air can be measured using direct methods or by indicator organisms.
- Direct methods include oxygen probes attached to computers that can measure the exact levels of oxygen in a pond. Special chemicals can be used to indicate levels of nitrate pollution from **fertilisers**.
- The presence or absence of an **indicator species** is used to estimate levels of pollution. For example:
  - The mayfly larva is an insect that can only live in clean water.
  - The water louse, bloodworm and mussels can live in polluted water.
  - Lichen grows on trees and rocks but only when the air is clean. It is unusual to find lichen growing in cities, because it is killed by the pollution from motor engines.

D–C

Remember!
If many different species are present in a habitat, this is usually a sign of low levels of pollution.

- There are advantages to the different methods of measuring pollution:
  - Using indicator organisms is cheaper, does not need equipment that can go wrong and monitors pollution levels over long periods of time.
  - Using direct methods can give more accurate results at any specific time.

B–A*

Improve your grade

Population and pollution

Explain the reasons for the increase in carbon dioxide levels in the atmosphere and explain why people are concerned about this. *AO1* [4 marks]

Sustainability

Conservation

D–C

- **Conservation** involves trying to preserve the variety of plants and animals and the **habitats** that they live in. People think that this is important because it can:
  - – protect our food supply
  - – prevent any damage to food chains, which can be hard to predict
  - – protect plants and animals that might be useful for medical uses
  - – protect organisms and habitats that people enjoy to visit and study.
- **Species** are at risk of **extinction** if the number of individuals or habitats falls below critical levels.

B–A*

- When trying to conserve species the important factors to bear in mind include:
  - – the size of the **population** (if the population is below a critical level there is unlikely to be enough genetic **variation** in the population to enable it to survive)
  - – the number of suitable habitats that are available for the organism to live in
  - – how much competition there is from other species.

Whale conservation

D–C

- Whales have been hunted for hundreds of years for their body parts, which are used in many products. Live whales are also important to the tourist trade.
- Some whales are kept in captivity for research or **captive breeding** programmes or just for our entertainment. However, many people object when whales lose their freedom.

Many uses of whale parts

B–A*

- It has been difficult to set up and police international agreements on whaling and some countries want to lift the ban on whaling.
- Scientists believe there is a need to kill some whales to help find out more about how whales can survive at extreme depths. However, they could study the whales without killing them. Migration patterns and whale communication can only be investigated if the animal is alive.

Sustainable development

D–C

- **Sustainable development** means taking enough resources from the environment for current needs, while leaving enough for the future and preventing permanent damage. For example:
  - – Fishing quotas are set, so that there are enough fish left to breed.
  - – Woods are replanted to keep up the supply of trees.

B–A*

- As the world population increases it is crucial to carry out sustainable development.
  - – **Fossil fuels** will run out. As there is an increase in demand for **energy**, we must manage alternative fuels, such as wood.
  - – We need to supply increasing amounts of food for growing populations without destroying large areas of natural habitats.
  - – Large amounts waste products must be **disposed** of so as to prevent or minimise pollution.
- If all of this is achieved it should help to save **endangered** species.

EXAM TIP

You should be able to suggest how an endangered organism could best be conserved.

Improve your grade

Saving endangered species

The Hawaiian goose is only found on the islands of Hawaii. In the mid-1900s only about 30 were left alive. The space on the islands is restricted and a number of animals have been introduced to the islands. Write about the problems facing scientists trying to save the goose from extinction. *AO2* [4 marks]

B2 Summary

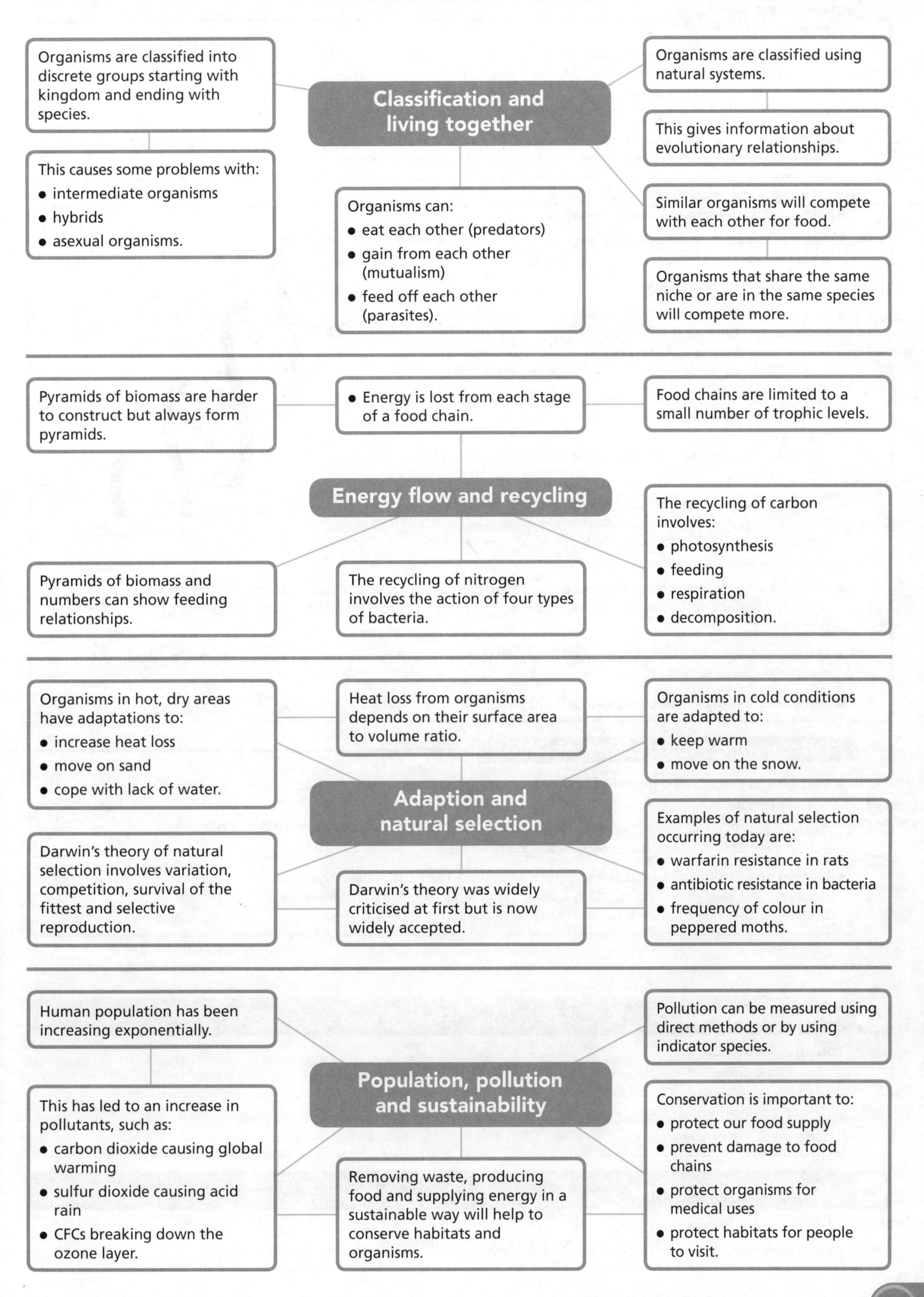

Organisms are classified into discrete groups starting with kingdom and ending with species.
This causes some problems with:
• intermediate organisms
• hybrids
• asexual organisms.
Classification and living together
Organisms can:
• eat each other (predators)
• gain from each other (mutualism)
• feed off each other (parasites).
Organisms are classified using natural systems.
This gives information about evolutionary relationships.
Similar organisms will compete with each other for food.
Organisms that share the same niche or are in the same species will compete more.
Pyramids of biomass are harder to construct but always form pyramids.
• Energy is lost from each stage of a food chain.
Food chains are limited to a small number of trophic levels.
Energy flow and recycling
Pyramids of biomass and numbers can show feeding relationships.
The recycling of nitrogen involves the action of four types of bacteria.
The recycling of carbon involves:
• photosynthesis
• feeding
• respiration
• decomposition.
Organisms in hot, dry areas have adaptations to:
• increase heat loss
• move on sand
• cope with lack of water.
Heat loss from organisms depends on their surface area to volume ratio.
Organisms in cold conditions are adapted to:
• keep warm
• move on the snow.
Adaption and natural selection
Darwin's theory of natural selection involves variation, competition, survival of the fittest and selective reproduction.
Darwin's theory was widely criticised at first but is now widely accepted.
Examples of natural selection occurring today are:
• warfarin resistance in rats
• antibiotic resistance in bacteria
• frequency of colour in peppered moths.
Human population has been increasing exponentially.
This has led to an increase in pollutants, such as:
• carbon dioxide causing global warming
• sulfur dioxide causing acid rain
• CFCs breaking down the ozone layer.
Population, pollution and sustainability
Removing waste, producing food and supplying energy in a sustainable way will help to conserve habitats and organisms.
Pollution can be measured using direct methods or by using indicator species.
Conservation is important to:
• protect our food supply
• prevent damage to food chains
• protect organisms for medical uses
• protect habitats for people to visit.

Molecules of life

Cell structure

D–C

- The number of **mitochondria** in the cytoplasm of a cell depends on the activity of the cell. This is because **respiration** occurs in mitochondria. Cells such as liver or muscle cells have large numbers of mitochondria. This is because the liver carries out many functions and muscle cells need to contract. Both types of cell therefore need lots of energy.

B–A*

- **Ribosomes** are smaller than mitochondria and are also found in the cytoplasm. They are too small to be seen with a light microscope and are the site of protein synthesis.

DNA and the genetic code

D–C

- The **nucleus** contains **genes**. Each gene:
  - is a section of a **chromosome** made of **DNA**
  - codes for a particular protein.
- DNA is made of two strands coiled to form a double helix, each strand containing chemicals called bases. There are four different types of bases, with cross links between the strands formed by pairs of bases. Each gene contains a different sequence of bases.
- Proteins are made in the cytoplasm but DNA cannot leave the nucleus. This means that a copy of the gene needs to be made that can leave the nucleus and carry the code to the cytoplasm.

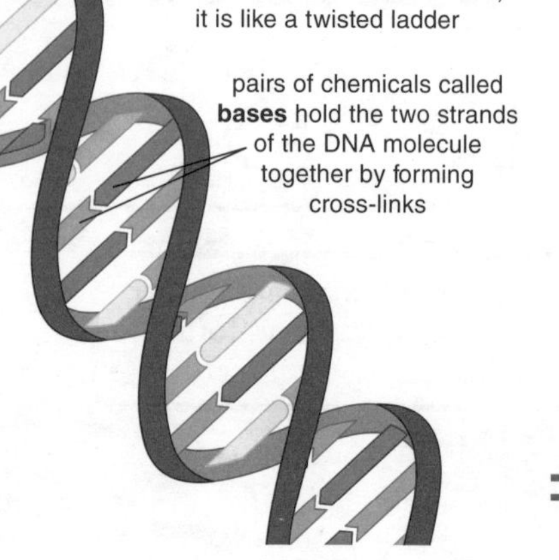

The structure of DNA

B–A*

- The four bases in DNA are called A, T, C and G. The cross links holding the two strands together are always between the same bases, A–T and G–C. This is called complementary base pairing.
- The **DNA base** code controls which protein is made. This is because the base sequence in the DNA codes for the **amino acid** sequence in the protein. Each amino acid is coded for by a sequence of three bases.
- The code needed to produce a protein is carried from the DNA to the ribosomes by a **molecule** called **messenger RNA**, or mRNA.
- Many of the proteins that are made are **enzymes,** which can control the activity of the cell.

Discovering the structure of DNA

D–C

- Watson and Crick built a model of DNA using data from other scientists. Two of the important pieces of data they used were:
  - photographs taken using **x-rays** which showed that DNA had two chains wound in a helix
  - data indicating that the bases occurred in pairs.

B–A*

- Watson and Crick worked out the structure of DNA in 1953 and shared the Nobel prize for this discovery in 1962. There is often such a delay between a discovery and the award of prizes because other scientists need to check the discovery to make sure that it is correct.

How science works

You should be able to:

- describe examples of how scientists made a series of observations in order to develop new scientific explanations
- understand that unexpected observations or results can lead to new scientific developments in the understanding of science.

Improve your grade

Protein synthesis

Describe where and how proteins are coded for and made. *AO1* [6 marks]

Proteins and mutations

Types of proteins

- All proteins are made of long chains of **amino acids** joined together.
- Proteins have different functions. Some examples are:
  - structural proteins used to build cells and tissues, e.g. **collagen**
  - **hormones**, which carry messages to control a reaction, e.g. **insulin** controls **blood sugar levels**
  - carrier proteins, e.g. **haemoglobin**, which carries oxygen
  - **enzymes**.
- Each protein has its own number and order of amino acids. This makes each type of protein **molecule** a different shape and gives it a different function.

Enzymes

- Enzymes speed up reactions in the body and so are called **biological catalysts**.
- They catalyse chemical reactions occurring in **respiration**, **photosynthesis** and protein synthesis of living cells.
- The **substrate** molecule fits into the **active site** of the enzyme like a key fitting into a lock:
  - This is why enzymes are described as working according to the 'lock and key mechanism'.
  - It also explains why each enzyme can only work on a particular substrate. This is called specificity and it happens because the substrate has to be the right shape.

The lock and key theory

- Enzymes all work best at a particular temperature and pH. This is called the optimum. Any change away from the optimum will slow down the reaction.
- Enzyme activity is affected by **pH** and temperature:
  - At low temperatures molecules are moving more slowly and so the enzyme and substrate are less likely to collide.
  - At very high or low pH values and at high temperatures the enzyme active site changes shape. This is called denaturing. The substrate cannot fit, so cannot react so quickly.
- It is possible to work out how temperature alters the **rate of reaction** by calculating the **temperature coefficient**, called Q_{10}. This is done for a 10 °C change in temperature, using:

$$Q_{10} = \frac{\text{rate at higher temperature}}{\text{rate at lower temperature}}$$

Mutations

- **Mutations** may occur spontaneously but can be made to occur more often by **radiation** or chemicals.
- When they occur, mutations:
  - may lead to the production of different proteins
  - are often harmful but may have no effect
  - occasionally they might give the individual an advantage.

EXAM TIP

A mutation changes the *order* of the four bases. Do not say that different bases are involved.

- Although every cell in the body has the same **genes** it does not mean that the all the same proteins are made. This is because different genes are switched off in different cells. This allows different cells to perform different functions.
- Gene mutations alter or prevent the production of the protein that is normally made, because they change the base code of **DNA**, and so change the order of amino acids in the protein.

Improve your grade

Mutations and enzymes

A mutation can occur in a gene that codes for an enzyme. Explain how a mutation could lead to an enzyme failing to work properly. *AO1* [6 marks]

Respiration

Why is respiration important?

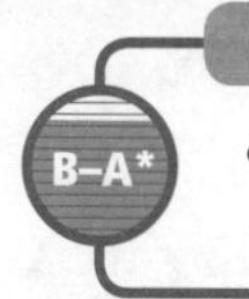

B–A*

- **Respiration** releases energy from food and this energy is trapped in a **molecule** called **ATP**. ATP can then be used to provide the energy for many different processes in living organisms.

Aerobic respiration

D–C

- **Aerobic respiration** involves the use of oxygen. The symbol equation for aerobic respiration is:

$$C_6H_{12}O_6 + 6O_2 \rightarrow 6CO_2 + 6H_2O$$

EXAM TIP

When you write this particular equation, make sure that all the letters are capitals and that the numbers are the right size and in the right position.

Anaerobic respiration

D–C

- During exercise, despite an increase in breathing rate and heart rate, the muscles often do not receive sufficient oxygen. They start to use **anaerobic respiration** in addition to aerobic respiration.
- The word equation for anaerobic respiration is:

  glucose → lactic acid (+ energy)
- Anaerobic respiration has two main disadvantages over aerobic respiration.
  - The lactic acid that is made by anaerobic respiration builds up in muscles, causing pain and fatigue.
  - Anaerobic respiration releases much less energy per glucose molecule than aerobic respiration.

B–A*

- The incomplete breakdown of glucose resulting in the build-up of lactic acid is called the **oxygen debt.**
- During recovery the breathing rate and heart rate stay high so that:
  - rapid blood flow can carry lactic acid away to the liver
  - extra oxygen can be supplied, enabling the liver to break down the lactic acid.

An athlete running long distances tries to use only aerobic respiration

Measuring respiration rate

D–C

- It is possible to set up different experiments to measure the rate of respiration. Two ways to do this involve:
  - measuring how much oxygen is used up – the faster it is consumed, the faster the respiration rate
  - the rate at which **carbon dioxide** is made.
- Scientists can use these results to calculate the **respiratory quotient.** This is worked out using the formula:

$$RQ = \frac{\text{carbon dioxide produced}}{\text{oxygen used}}$$

Remember!

The RQ for glucose is 1, from the equation:

$$\frac{6CO_2}{6O_2} = 1$$

B–A*

- The **metabolic rate** is described as the sum of all the reactions that are occurring in the body. If the metabolic rate is high, more oxygen is needed, as aerobic respiration is faster.
- Changes in temperature and **pH** can also change the respiration rate because they affect **enzymes**, and respiration is controlled by enzymes.

Improve your grade

Respiration and enzymes

An experiment was set up to measure the oxygen uptake of insects. The data shows that the maximum uptake was at about 35 °C. Above and below that temperature less oxygen was used.

Explain what the oxygen is used for and why it is used fastest at 35 °C. *AO1* [2 marks], *AO2* [2 marks]

Cell division

Becoming multicellular

- There are a number of advantages of being **multicellular**, as humans are. It allows an organism to become larger and more complex. It also allows different cells to take on different jobs. This is called **cell differentiation**.
- However, when an organism becomes multicellular, it needs to have systems that can:
  - allow communication between all the cells in the body
  - supply all the cells with enough nutrients
  - control exchanges with the environment such as heat and gases.

Mitosis

- The process that produces new cells for growth is called **mitosis**.
- The cells that are made by mitosis are genetically identical. Before cells divide, **DNA** replication must take place. This is so that each cell produced still has two copies of each **chromosome**.
- Body cells in **mammals** have two copies of each chromosome, so they are called **diploid** cells.
- Before mitosis happens, DNA is replicated. This involves:
  - the two strands of the DNA **molecule** 'unzipping' to form single strands
  - new double strands forming by **DNA bases** lining up in complementary pairings.
- Then mitosis occurs. The chromosomes line up along the centre of the cell and divide. The copies then move to opposite poles (ends) of the cell.

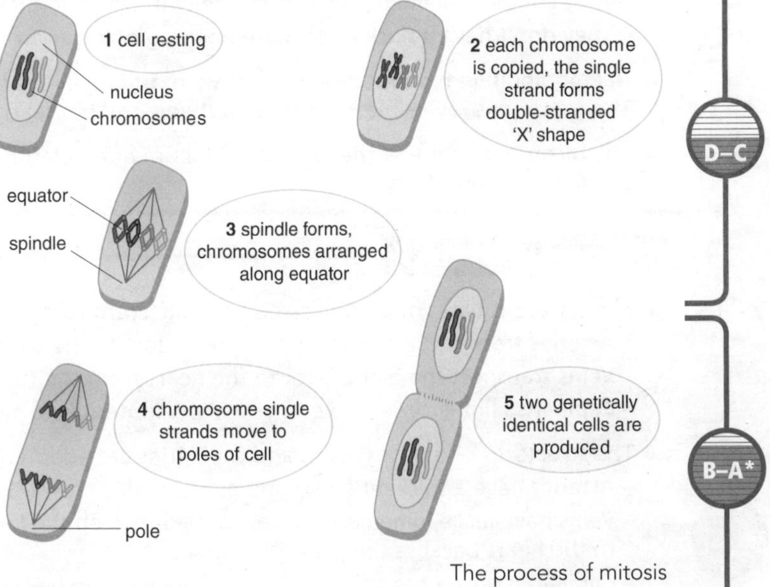

The process of mitosis

Meiosis

- The type of cell division that produces **gametes** is called **meiosis**.
- Gametes are **haploid** cells because they contain only one chromosome from each pair. This means that the zygote gets one copy of a **gene** from one parent and another copy from the other parent. This produces genetic **variation**.
- The structure of a sperm cell is adapted to its function. It has:
  - many **mitochondria** to provide the energy for swimming to the egg
  - an **acrosome** that releases **enzymes** to digest the egg membrane.

diploid cells – the single strands are copied to make x-shaped chromosomes

1 chromosomes pair up

pole

2 one from each pair moves to each pole

3 the strands of each chromosome are pulled apart to opposite poles

4 four new haploid cells form, all genetically different from each other

The process of meiosis

- In meiosis, there are two divisions.
  First, the single strands are copied to make X-shaped chromosomes and chromosomes with the same genes pair up. Then:
  - In the first division, one chromosome from each pair moves to opposite poles of the cell.
  - In the second division, the copies of each chromosome come apart and move to opposite poles of the cell.

Improve your grade

Meiosis and mitosis

The two processes mitosis and meiosis occur in the human body. Compare each process, writing about where they occur and any differences in the process. *AO1* [6 marks]

EXAM TIP

Be careful not to mix up mitosis and meiosis.

The circulatory system

Blood

D–C

- The liquid part of the blood called **plasma** carries a number of important substances around the body:
  - dissolved food substances such as glucose
  - **carbon dioxide** from the tissues to the lungs
  - **hormones** from the glands where they are made, to their target cells
  - plasma proteins such as **antibodies**
  - waste substances such as urea.
- **Red blood cells** are adapted to their function of carrying oxygen in a number of ways:
  - They are very small so that they can pass through the smallest blood vessels.
  - They are shaped like biconcave discs so that they have a large surface area to exchange oxygen quicker.
  - They contain **haemoglobin** to combine with oxygen. It is the haemoglobin that makes them appear red.
  - They don't have a **nucleus** so more haemoglobin can fit in.

B–A*

- Haemoglobin in red blood cells reacts with oxygen in the lungs, forming oxyhaemoglobin. The reaction is reversible: when the oxyhaemoglobin reaches the tissues, the oxygen is released.
- The biconcave shape of the red blood cell provides a larger surface area to volume ratio to exchange oxygen more quickly.

Blood vessels

D–C

- The different types of blood vessels have different jobs:
  - **Arteries** transport blood away from the heart to the tissues.
  - **Veins** transport the blood back to the heart from the tissues.
  - **Capillaries** link arteries to veins and allow materials to pass between the blood and the tissues.

B–A*

- The structures of arteries, veins and capillaries are adapted to carry out specific functions:
  - Arteries have a thick muscular and elastic wall to resist the high pressure.
  - Veins have large lumen and valves to try and keep the blood moving back to the heart because the pressure is low.
  - Capillaries have permeable walls so substances can be transferred between the blood and the tissues.

Remember!
The blood in the pulmonary vein is oxygenated and in the pulmonary artery it is deoxygenated, unlike in other veins and arteries.

The heart

D–C

- The different parts of the heart work together to circulate the blood.
  - The left and right atria receive blood from veins.
  - The left and right ventricles pump blood out into arteries.
  - The semilunar, tricuspid and bicuspid valves prevent any backflow of blood.
  - The pulmonary veins and the vena cava are the main veins carrying blood back to the heart.
  - The aorta and pulmonary arteries carry blood away from the heart.
- The left ventricle has a thicker muscle wall than the right ventricle because it has to pump blood all round the body rather than just to the lungs, which are close by.

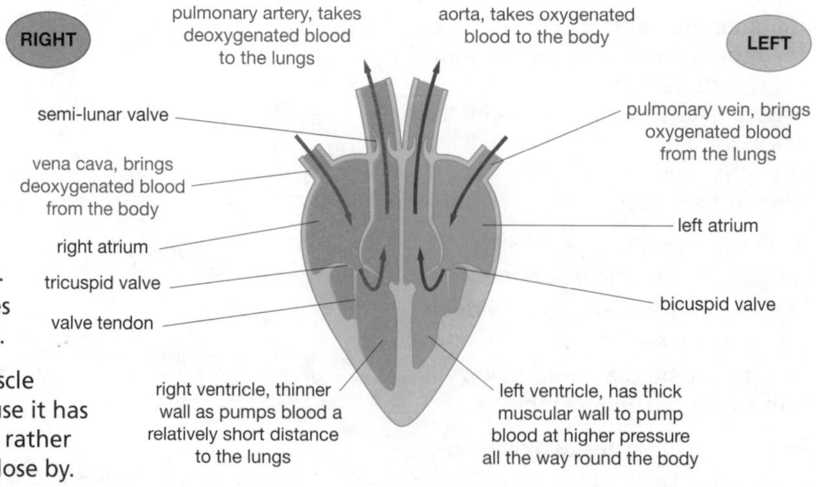

RIGHT
pulmonary artery, takes deoxygenated blood to the lungs
aorta, takes oxygenated blood to the body
LEFT
semi-lunar valve
vena cava, brings deoxygenated blood from the body
right atrium
tricuspid valve
valve tendon
pulmonary vein, brings oxygenated blood from the lungs
left atrium
bicuspid valve
right ventricle, thinner wall as pumps blood a relatively short distance to the lungs
left ventricle, has thick muscular wall to pump blood at higher pressure all the way round the body

Blood flow in the heart

B–A*

- The blood is pumped to the lungs and returns to the heart to be pumped to the body. This is called a **double circulatory system**. This means the blood is at a higher pressure and so flows to the tissues at a faster rate.

Improve your grade

Valves

Valves are found in veins, at the start of the arteries leaving the heart and in the heart. Write about the importance of these different valves. *AO1/2* [2 marks]

Growth and development

Different types of cells

D–C

- Bacterial cells differ from plant and animal cells in that they lack a 'true' **nucleus, mitochondria** and chloroplasts.

B–A*

- As bacterial cells do not have a true nucleus, **DNA** is found in the cytoplasm as a single circular strand or **chromosome**.

Measuring growth

D–C

- The diagram of a typical growth curve shows the main phases of growth.
- Two of these phases involve rapid growth; one is just after birth and the other in adolescence.
- Dry mass is the best measure of growth.

B–A*

- Measuring growth by length is easy to do but only measures growth in one direction.
- Measuring wet mass is hard to do for some organisms, e.g. trees, but is easy for animals. However, the water content of organisms can vary with time.
- Dry mass can only be measured by killing the organism and driving off the water but it does measure the true growth of the whole organism.
- Different parts of an organism may grow at different rates compared to the whole organism. This is because different parts of the organism may be needed at different times during the life of the organism.

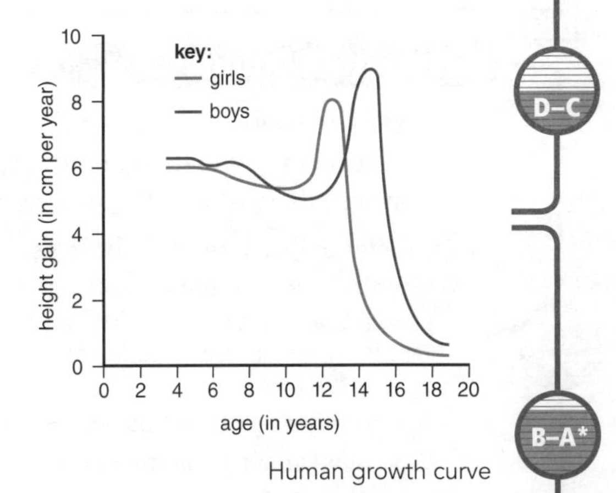

Human growth curve

Differentiation

D–C

- Cells called **stem cells** stay undifferentiated. They can develop into different types of cells.
- Stem cells can be obtained from embryos and could potentially be used to treat some medical conditions.
- However, there are issues arising from stem cell research in animals. Some people think that it is wrong because the embryos are destroyed. Others think that this is acceptable as it can treat life-threatening diseases.

B–A*

- Stem cells can be found in the adult as well as in the embryo. Embryonic stem cells can form a greater range of cell types and are easier to find.

EXAM TIP

Make sure that you can give both sides of this argument, even if you have strong views.

Plant and animal growth

D–C

- Differences between plant and animal growth include the following.
  - Animals tend to only grow to a certain size but many plants can carry on growing.
  - Plant cell division only happens in areas called **meristems**, found at the tips of roots and shoots.
  - The main way that plants gain height is by cells enlarging rather than dividing.
  - Many plant cells keep the ability to differentiate, but most animal cells lose it at an early stage.

Improve your grade

A plant growth curve

Katie wants to plot a growth curve for a broad bean plant using dry mass. Given 100 broad bean seeds, explain how you would collect the data to plot the graph and explain why she wants to plot dry mass. *AO1* [5 marks]

New genes for old

Selective breeding

D–C

- There are possible problems with **selective breeding** programmes. They may lead to **inbreeding**, where two closely related individuals mate, and this can cause health problems within the **species**. Certain breeds of dog show these problems.

B–A*

- Inbreeding can reduce the variety of **alleles** in the population (the **gene pool**). This can lead to:
  - an increased risk of harmful recessive characteristics showing up in offspring
  - a reduction in **variation**, so that **populations** cannot adapt to change so easily.

Genetic engineering

D–C

- **Genetic engineering** has advantages and risks:
  - One advantage is that organisms with desired features can be produced very quickly.
  - However, there is a risk that the inserted genes may have unexpected harmful side effects.
- Examples of organisms that have been made using genetic engineering include the following:
  - Rice that contains beta-carotene has been made by inserting the genes that control beta-carotene production from carrots. Humans can then convert the beta-carotene from rice into Vitamin A. This is important because in some parts of the world people rely on rice, which normally has very little Vitamin A.
  - Genetically engineered **bacteria** have been made that produce human **insulin**.
  - Crop plants have been made that are resistant to **herbicides**, frost damage or disease.
- A number of ethical issues are involved in genetic engineering:
  - Some people are worried about possible long-term side effects, e.g. that genetically engineered plants or animals will disturb natural ecosystems.
  - Other people think that it is morally wrong, whatever the intended benefits.

B–A*

- To carry out genetic engineering, four steps are taken:
  - The desired characteristics are selected.
  - The genes responsible are identified and removed (isolation).
  - The genes are inserted into other organisms.
  - The organisms are allowed to reproduce (replication).

Remember!

It is the ***gene*** that is put into another organism, e.g. the gene for human insulin, not human insulin itself.

Gene therapy

D–C

- The process of using genetic engineering to change a person's genes and cure certain disorders is called **gene therapy**.

B–A*

- Gene therapy could involve body cells or **gametes**. Changing the genes in gametes is much more controversial. This is because it is sometimes difficult to decide which genes parents should be allowed to change. It could lead to 'designer babies'.

Improve your grade

Spider silk

Spider silk is very strong and could be very useful in industry. Goats have now been produced that make spider silk in their milk.

Describe how this could be done and suggest reasons why this method of production might be more useful. *AO2* [4 marks]

Cloning

Cloning animals

- Dolly the sheep was produced by a process called **nuclear transfer.** This involves removing the **nucleus** from a body cell and placing it into an egg cell that has had its nucleus removed.
- Animals could be **cloned** to:
  - mass-produce animals with desirable characteristics
  - produce animals that have been genetically engineered to provide human products
  - produce human embryos to supply **stem cells** for therapy.
- There are some ethical dilemmas concerning human cloning. Some people think that it is wrong to clone people as they will not be 'true individuals'.

D–C

egg cell taken from sheep A and nucleus removed

cells taken from the udder of sheep B and the nucleus removed

nucleus from sheep B is put into egg of sheep A

after being given an electric shock to make it divide, the egg cell is put into a surrogate mother sheep to grow

Nuclear transfer

- When Dolly was produced by nuclear transfer, it involved a number of steps:
  - The donor egg cell had its nucleus removed.
  - The egg cell nucleus was replaced with the nucleus from an udder cell.
  - The egg cell was then given an electric shock to make it divide.
  - The embryo was implanted into a surrogate mother sheep.
  - The embryo then grew into a clone of the sheep from which the udder cell came.
- Cloning technology could be useful in a number of ways. Some of these may carry risks. For example, genetically modified animals could be cloned to supply replacement organs for humans. Some people are worried that this could lead to diseases being spread from animals to humans.

Remember!
Cloning would not be very successful in recreating endangered or extinct animals, as all the clones would be the same sex and genetically identical.

B–A*

Cloning plants

- Producing cloned plants has advantages and disadvantages.
  - Advantages: growers can be sure of the characteristics of each plant since all the plants will be genetically identical. It is also possible to mass-produce plants that may be difficult to grow from seed.
  - Disadvantages: if the plants become susceptible to disease or to change in the environmental conditions, then all the plants will be affected. There is a lack of genetic **variation** in the plants.

D–C

- Plants can be cloned by a process called **tissue culture,** as follows:
  - A plant is selected that has certain characteristics.
  - A large number of small pieces of tissue are then cut from the plant.
  - The small pieces of tissue are grown in test tubes or dishes containing a growth medium.
  - Aseptic technique is used at all stages to stop any **microbes** infecting the plants.
- Cloning plants is easier than cloning animals because many plant cells retain the ability to differentiate. Animal cells, however, usually lose this ability at an early stage.

B–A*

Improve your grade

Cloning plants

A garden centre wants to sell an attractively coloured geranium plant. They decide to produce many clones of the plant using tissue culture.

(a) What are the possible disadvantages of this method of reproduction? *AO1* [2 marks]

(b) Why is this method not possible for producing goldfish for garden ponds? *AO2* [2 marks]

B3 Summary

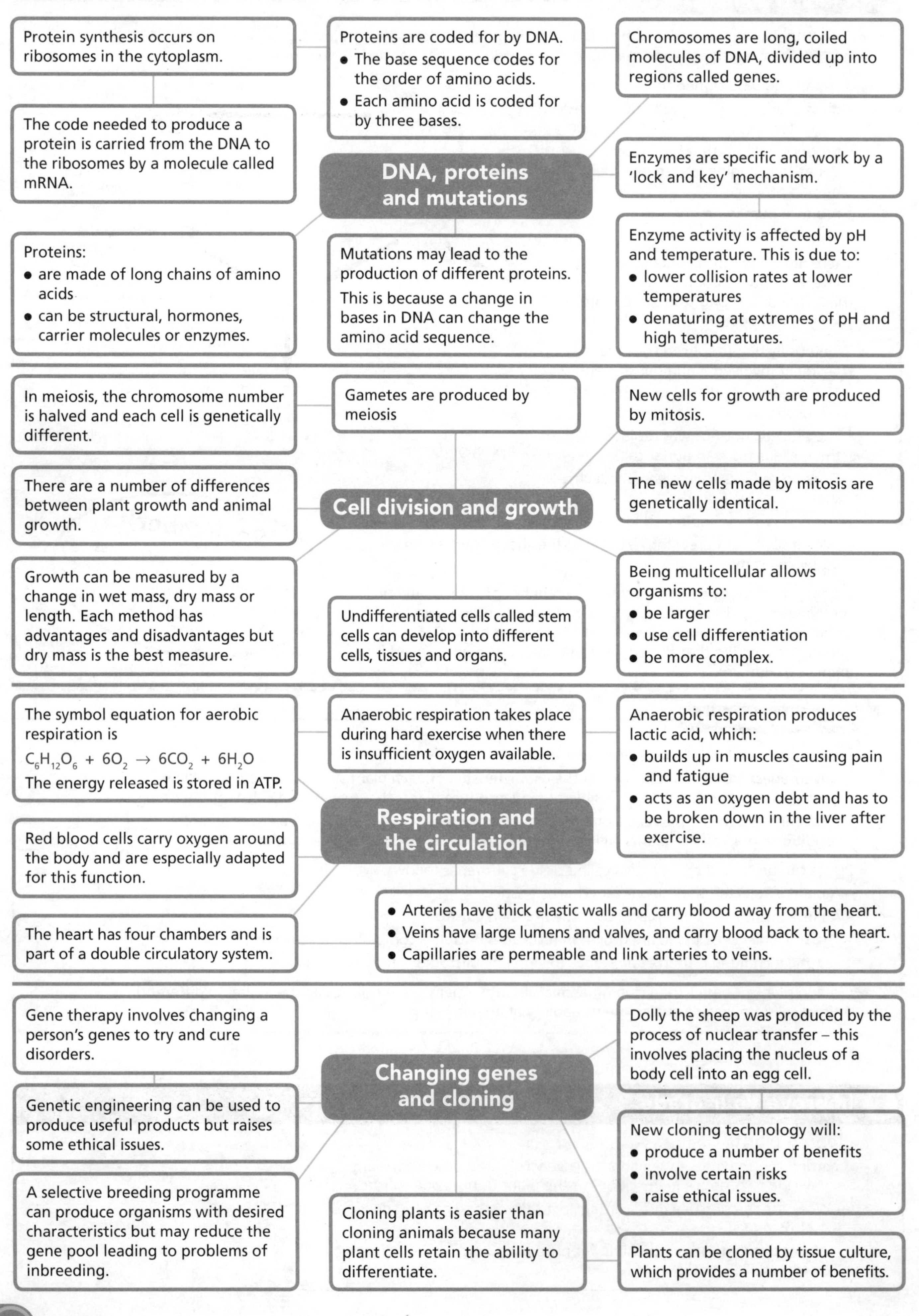

Protein synthesis occurs on ribosomes in the cytoplasm.
Proteins are coded for by DNA.
• The base sequence codes for the order of amino acids.
• Each amino acid is coded for by three bases.
Chromosomes are long, coiled molecules of DNA, divided up into regions called genes.
The code needed to produce a protein is carried from the DNA to the ribosomes by a molecule called mRNA.
DNA, proteins and mutations
Enzymes are specific and work by a 'lock and key' mechanism.
Proteins:
• are made of long chains of amino acids
• can be structural, hormones, carrier molecules or enzymes.
Mutations may lead to the production of different proteins.
This is because a change in bases in DNA can change the amino acid sequence.
Enzyme activity is affected by pH and temperature. This is due to:
• lower collision rates at lower temperatures
• denaturing at extremes of pH and high temperatures.
In meiosis, the chromosome number is halved and each cell is genetically different.
Gametes are produced by meiosis
New cells for growth are produced by mitosis.
There are a number of differences between plant growth and animal growth.
Cell division and growth
The new cells made by mitosis are genetically identical.
Growth can be measured by a change in wet mass, dry mass or length. Each method has advantages and disadvantages but dry mass is the best measure.
Undifferentiated cells called stem cells can develop into different cells, tissues and organs.
Being multicellular allows organisms to:
• be larger
• use cell differentiation
• be more complex.
The symbol equation for aerobic respiration is
$C_6H_{12}O_6 + 6O_2 \rightarrow 6CO_2 + 6H_2O$
The energy released is stored in ATP.
Anaerobic respiration takes place during hard exercise when there is insufficient oxygen available.
Anaerobic respiration produces lactic acid, which:
• builds up in muscles causing pain and fatigue
• acts as an oxygen debt and has to be broken down in the liver after exercise.
Respiration and the circulation
Red blood cells carry oxygen around the body and are especially adapted for this function.
• Arteries have thick elastic walls and carry blood away from the heart.
• Veins have large lumens and valves, and carry blood back to the heart.
• Capillaries are permeable and link arteries to veins.
The heart has four chambers and is part of a double circulatory system.
Gene therapy involves changing a person's genes to try and cure disorders.
Dolly the sheep was produced by the process of nuclear transfer – this involves placing the nucleus of a body cell into an egg cell.
Changing genes and cloning
Genetic engineering can be used to produce useful products but raises some ethical issues.
New cloning technology will:
• produce a number of benefits
• involve certain risks
• raise ethical issues.
A selective breeding programme can produce organisms with desired characteristics but may reduce the gene pool leading to problems of inbreeding.
Cloning plants is easier than cloning animals because many plant cells retain the ability to differentiate.
Plants can be cloned by tissue culture, which provides a number of benefits.

Ecology in the local environment

Distribution of organisms

- An ecosystem, such as a garden, is made up of all the plants and animals living there and their surroundings. Where a plant or animal lives is its **habitat**.
- All the animals and plants living in the garden make up the **community**. The number of a particular plant or animal present in the community is called its **population.**
- Natural ecosystems, such as native woodlands and lakes, have a large variety of plants and animals living there – this means it has good **biodiversity**. Artificial ecosystems, such as forestry plantations and fish farms, have poor biodiversity.
- The distribution of organisms can be mapped using a **transect** line. A long length of string is laid across an area such as a path or sea shore. At regular intervals the organisms in a square frame called a quadrat can be counted (for animals) or assessed for percentage cover (for plants). The data can be displayed as a **kite diagram**.

Remember!
Artificial ecosystems are created by humans, for the benefit of humans

D–C

- In artificial ecosystems, humans deliberately keep and protect only one **species** (such as salmon in a fish farm) and remove any other organisms that would compete with it and lower the yield. This does not happen in a natural ecosystem.
- A transect line can show zonation in the distribution of organisms. Changes in abiotic (not biological) factors such as exposure on a sea shore or trampling near a footpath, cause zonation.
- Food chains and food webs show that plants and animals are interdependent, with energy being transferred from one organism to another. The exchange of gases in **photosynthesis** and **respiration** ensures an overall balance of these gases. An ecosystem is therefore self-supporting in all factors apart from having to have the Sun as an energy source.

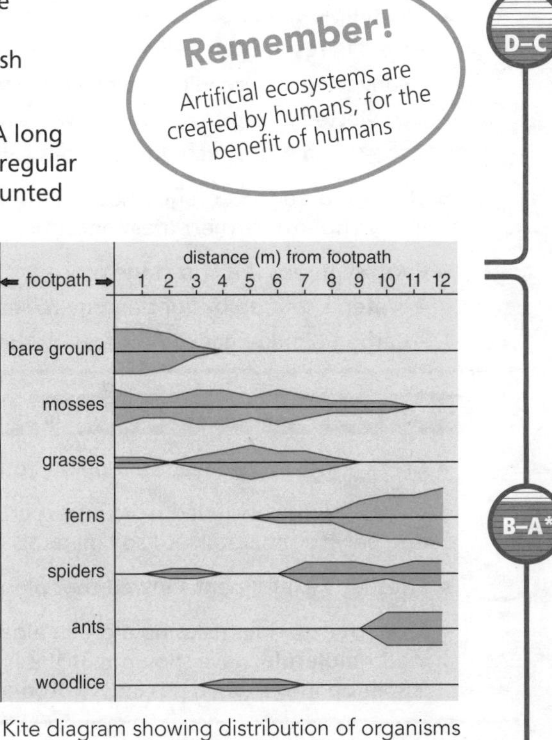

Kite diagram showing distribution of organisms near a path

B–A*

Population size

- Population size can be estimated by obtaining data from a small sample and scaling up. For example, Shabeena has a large lawn with some dandelions in it. She counts the number of dandelions in ten 1 m^2 quadrats and calculates the mean as 8 dandelions/1 m^2. Her lawn is 100 m^2. She calculates that there are 800 dandelions in her lawn.
- Shabeena also wants to estimate the woodlouse population in her garden. She uses a capture–recapture method.

$$\text{Population size} = \frac{\text{number in 1st sample} \times \text{number in 2nd sample}}{\text{number in 2nd sample previously marked}}$$

- Shabeena captures 36 woodlice in a pitfall trap, marks them with a white dot and releases them. Later she captures 48 woodlice, 6 with a white dot.

$$\text{Population size} = \frac{36 \times 48}{6} = 288 \text{ woodlice}$$

D–C

- Shabeena realises that if she had used a bigger quadrat and more samples, her estimation would have been more accurate.
- Using a capture–recapture method assumes:
  – There are no deaths or reproduction and no movement of animals into and out of the area.
  – Identical sampling methods are used for both samples.
  – The markings do not affect the survival of the woodlice.

B–A*

Improve your grade

Distribution of animals

Rick wants to find out about the zonation of different types of limpets down a sea shore. Explain:

(a) how he should do this
(b) what can cause this zonation.
AO1/2 [5 marks]

Photosynthesis

The chemistry of photosynthesis

D–C

- The **balanced symbol equation** for **photosynthesis** is:

$$6CO_2 + 6H_2O \xrightarrow[\text{chlorophyll}]{\text{light energy}} C_6H_{12}O_6 + 6O_2$$

EXAM TIP

It is important to write the correct chemical formula (e.g. CO_2, not CO2).

- The simple sugars such as glucose can be:
  - used in **respiration**, releasing energy
  - converted into cellulose to make cell walls
  - converted into proteins for growth and repair
  - converted into starch, fats and oils for storage.

B–A*

- Starch is used for storage since it is insoluble and does not move from storage areas. Unlike glucose, it does not affect the water concentration of cells and cause **osmosis**.
- Photosynthesis is a two-stage process.
  - Water is split up by light energy releasing oxygen gas and hydrogen **ions**.
  - **Carbon dioxide** gas combines with the hydrogen ions producing glucose and water.

Historical understanding of photosynthesis

D–C

- Greek scientists believed that plants took **minerals** out of the soil to grow and gain mass.
- Van Helmont concluded from his experiment on growing a willow tree that plant growth could not be due only to the uptake of soil minerals – it must depend on something else.
- Priestley's experiment showed that plants produce oxygen.

B–A*

- Modern experiments using a green alga called *Chlorella* and an **isotope** of oxygen, ^{18}O, as part of a water **molecule**, have shown that the light energy is used to split water, not carbon dioxide. The water is split up into oxygen gas and hydrogen ions. Isotopes are different forms of the same element.

The rate of photosynthesis

D–C

Remember!

Plants respire at all times, not just at night.

- The rate of photosynthesis can be increased by the plant having:
  - more carbon dioxide
  - more light
  - a higher temperature which increases **enzyme** action.
- Photosynthesis will only take place during daytime (in the light). However, since plants are living organisms they must also respire, releasing energy at all times.

B–A*

- Plants respire at all times by taking in oxygen and releasing carbon dioxide.

During the day, when it is light, they also carry out photosynthesis, taking in carbon dioxide and releasing oxygen: the same gas exchange as respiration but in reverse. The rate of gas exchange in photosynthesis is more than that of respiration in terms of quantities, so respiration can only be noticed at night (in darkness).

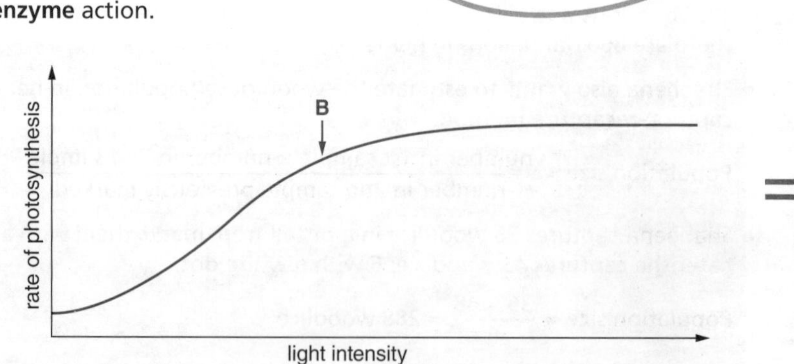

Effect of light on the rate of photosynthesis – at B, the rate of photosynthesis is being limited either by carbon dioxide concentration or temperature

- Since photosynthesis depends on light, temperature and carbon dioxide, a lack of one of these factors will limit the rate of photosynthesis. They are called limiting factors.

Improve your grade

Exchange of gases in plants

When gas exchange in plants is analysed, they seem to respire only at night. Explain why.
AO2 [4 marks]

Leaves and photosynthesis

Leaf structure

- A green leaf has many specialised cells, as shown in the diagram.

Specialised cells in a leaf

- The cells are adapted for efficient **photosynthesis**.
  - The outer epidermis lacks chloroplasts and so is transparent; there are no barriers to the entry of light.
  - The upper **palisade** layer contains most of the leaf's chloroplasts, as they will receive most of the light.
  - The **spongy mesophyll cells** are loosely spaced so that **diffusion** of gases between cells and the outside atmosphere can take place.
  - The arrangement of mesophyll cells creates a large surface area/volume ratio so that large amounts of gases can enter and exit the cells.

Leaf adaptations for photosynthesis

- Leaves are adapted so that photosynthesis is very efficient.
  - They are usually broad so that they have a large surface area to get as much light as possible.
  - They are usually thin so that gases can **diffuse** through easily and light can get to all cells.
  - They contain **chlorophyll** and other pigments so that they can use light from a broad range of the light spectrum.
  - They have a network of **vascular bundles (veins)** for support and transport of chemicals such as water and glucose.
  - They have specialised guard cells which control the opening and closing of **stomata** therefore regulating the flow of **carbon dioxide** and oxygen as well as water loss.
- By having many pigments (chlorophyll a and b, **carotene** and **xanthophylls**) the plant cells can maximise the use of the Sun's energy. Each pigment absorbs light of different wavelengths.

Remember!
The small holes in the lower epidermis are called stomata; they are surrounded by two guard cells.

Improve your grade

Absorption of light

Look at the graph.

(a) Which parts of the light spectrum are i) reflected and ii) used by leaves? *AO2* [2 marks]

(b) Explain why a leaf has many different pigments. *AO1* [2 marks]

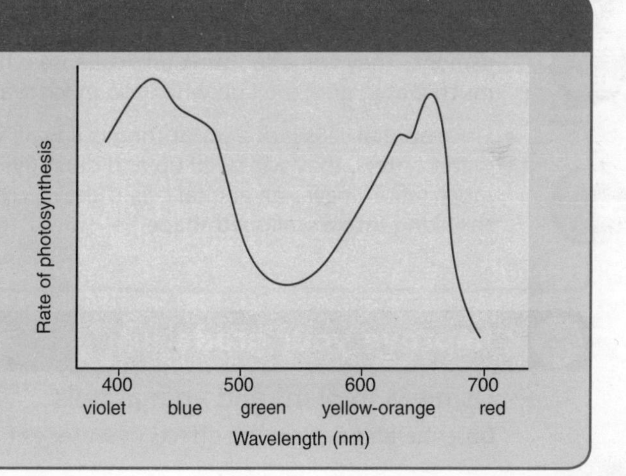

Diffusion and osmosis

Diffusion

D–C

- **Diffusion** is the net movement of particles in a gas or liquid from an area of high concentration to an area of low concentration, resulting from the **random** movement of the particles.
- This explains how **molecules** of water, oxygen and **carbon dioxide** can enter and leave cells through the **cell membrane**. If a plant cell is using up carbon dioxide, there is a lower concentration of it inside the cell, so carbon dioxide will enter by diffusion.
- Leaves are adapted to increase the rate of diffusion of carbon dioxide and oxygen by having:
  - (usually) a large surface area
  - specialised openings called **stomata**, which are spaced out
  - gaps between the **spongy mesophyll cells**.

B–A*

- The rate of diffusion is not a fixed quantity. The rate can be increased by having:
  - a shorter distance for the molecules to travel
  - a steeper concentration gradient (a greater difference in concentration between the two areas)
  - a greater surface area for the molecules to **diffuse** from, or into.

Osmosis

D–C

- **Osmosis** is a type of diffusion; it depends on the presence of a **partially-permeable membrane** that allows the passage of water molecules but not large molecules like glucose.
- Osmosis is the movement of water across a partially-permeable membrane from an area of high water concentration (a dilute solution) to an area of low water concentration (a concentrated solution).

B–A*

- Osmosis is a consequence of the random movement of water molecules, which is not restricted by a partially-permeable membrane. The net movement of water molecules will be from an area where there are many to one where there are few.
- Knowing the different concentrations of water inside and outside cells makes it possible to predict the net movement of water molecules.

Remember!
Osmosis requires the presence of a partially-permeable cell membrane.

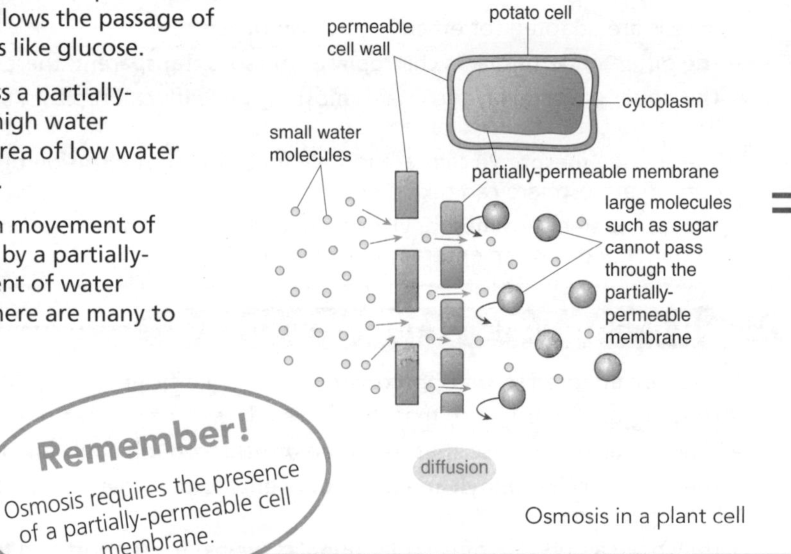

Osmosis in a plant cell

Water in cells

D–C

- The entry of water into plant cells increases the pressure pushing on the cell wall, which is rigid and not elastic. This **turgor pressure** supports the cell, stopping it, and the whole plant, from collapsing. When too much water leaves a cell, it loses this pressure and the plant wilts.

B–A*

- A plant cell full of water is said to be **turgid**. When the cell loses water the cell contents shrink and become **plasmolysed** and the cell is called **flaccid**.

D–C

- Animal cells also react to intake and loss of water due to osmosis. They will also shrink and collapse when they lose too much water, and swell up when too much water enters.

B–A*

- Since animal cells lack a supporting cell wall, when too much water enters, they will swell up and burst (**lysis**). When too much water leaves an animal cell, it shows **crenation** by shrinking into a scalloped shape.

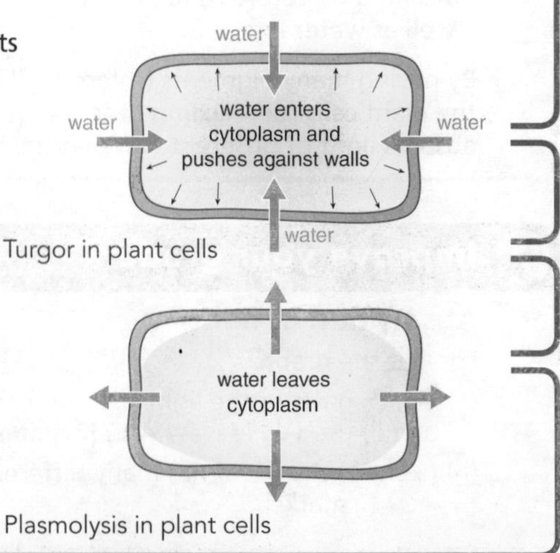

Turgor in plant cells

Plasmolysis in plant cells

Improve your grade

Osmosis in plant and animal cells

Describe and explain the effects of water entry into plant and animal cells. *AO1/2* [5 marks]

Transport in plants

Xylem and phloem cells

- **Xylem** and **phloem** are made up of specialised plant cells. Both types of tissue are continuous from the root, through the stem and into the leaf.
- Xylem and phloem cells form **vascular bundles** in dicotyledonous (broad-leaved) plants.
- Xylem cells carry water and **minerals** from the roots to the leaves and are therefore involved in transpiration. Phloem cells carry food substances such as sugars up and down stems to growing and storage tissues. This transport of food substances is called translocation.

D–C

EXAM TIP

Don't get confused between phloem (which carries food) and xylem (which carries water).

- The xylem cells are called vessels. They are dead cells, and the lack of living cytoplasm leaves a hollow lumen. Their cellulose walls have extra thickening of lignin, giving great strength and support.
- Phloem cells are living cells and are arranged in columns.

B–A*

Remember!

Xylem cells are involved in transpiration, phloem cells are involved in translocation.

Transpiration

- Transpiration is the **evaporation** (changing from a liquid into a gas) and **diffusion** of water from inside leaves. This loss of water from leaves helps to create a continuous flow of water from the roots to the leaves in xylem cells.
- Root hairs are projections from root hair cells. They produce a large surface area for water uptake by **osmosis**.
- Transpiration ensures that plants have water for cooling by evaporation, **photosynthesis** and support from cells' **turgor pressure**, and for transport of minerals.
- The rate of transpiration is increased by an increase in light intensity, temperature and air movement and a decrease in humidity (the amount of water vapour in the atmosphere).
- The structure of a leaf is adapted to prevent too much water loss, which could cause wilting. Water loss is reduced by having a waxy cuticle covering the outer epidermal cells and by most **stomatal** openings being situated on the shaded lower surface.

D–C

- Plant leaves are adapted for efficient photosynthesis by having stomata for entry and exit of gases. The **spongy mesophyll cells** are also covered with a film of water in which the gases can dissolve. This water can therefore readily escape through the stomata.
- The rate of transpiration can be increased by:
  - an increase in light intensity, which results in stomata being open
  - an increase in temperature, causing an increase in the evaporation of water
  - an increase in air movement, blowing away air containing a lot of evaporated water
  - a decrease in humidity (the amount of water vapour in the atmosphere), allowing more water to evaporate.

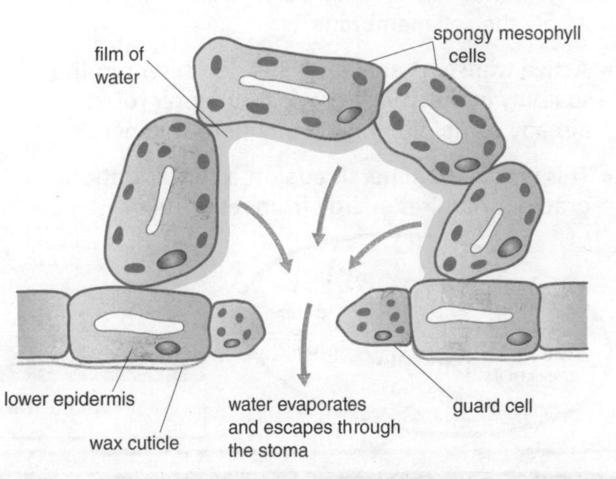

Movement of water through a stoma

- The structure of a leaf is adapted to reduce water loss. Its guard cells are able to change the size of the stomatal openings. The guard cells contain chloroplasts, so photosynthesis (in the presence of water and light) will produce sugars, increasing turgor pressure, causing the cells to swell. Due to differential thickness of their walls, the guard cells curve, opening the stoma.
- Further reduction in water loss is brought about by having fewer stomata, smaller stomata, the position of stomata (mainly in the lower epidermis), and their distribution.

B–A*

Improve your grade

Transport in plants

Marram grass grows in exposed sand dunes. It has narrow leaves, stomata sunk in pits, many hairs on its leaves and leaves that can curl up into a tube. Explain how these adaptations help it to survive. *AO2* [4 marks]

Plants need minerals

Use of minerals

- Plants need **minerals**, such as:
  - nitrates, to make proteins, which plants use for **cell** growth
  - phosphates, which are involved in **respiration** and growth
  - potassium compounds, which are involved in respiration and **photosynthesis**
  - magnesium compounds, which are involved in photosynthesis.
- Elements from soil minerals are used to produce useful compounds.
  - Nitrogen (from nitrates) is used to produce **amino acids**, which combine to form a variety of proteins.
  - Phosphorus (from phosphates), is used to make **DNA**, which contains the plant's genetic code, and cell membranes.
  - Potassium (from potassium compounds) is used to help **enzyme** action in photosynthesis and respiration; enzymes speed up chemical reactions.
  - Magnesium (from magnesium compounds) is used to make **chlorophyll**, which is essential for photosynthesis.

Mineral deficiency

- The lack of certain minerals results in specific symptoms:
  - Lack of nitrate causes poor growth and yellow leaves.
  - Lack of phosphate causes poor root growth and discoloured leaves.
  - Lack of potassium causes poor flower and root growth and discoloured leaves.
  - Lack of magnesium causes yellow leaves.

Mineral uptake

- Minerals are usually present in soil in low concentrations.
- Minerals are taken up by root hair cells by active transport, rather than by **diffusion** or **osmosis**. A system of carriers transport selected minerals across the cell **membrane**.
- **Active transport** enables minerals, present in the soil only in low concentrations, to enter root hairs already containing higher amounts of minerals.
- This uptake of minerals against a concentration gradient requires energy from respiration.

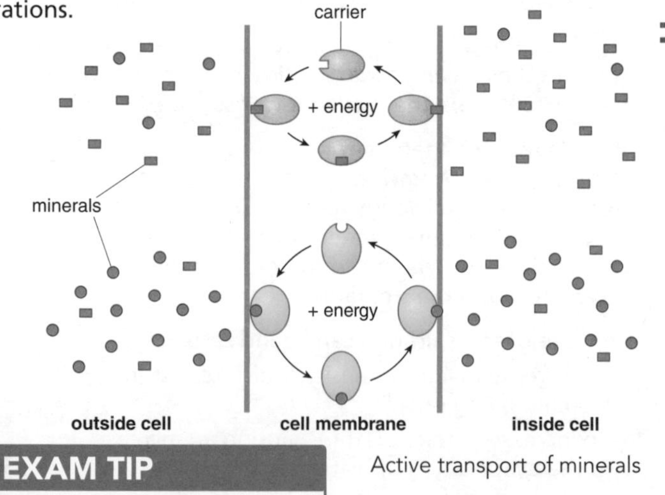

Active transport of minerals

Remember!
Mineral uptake involves active transport, rather than diffusion or osmosis.

EXAM TIP
Root *hairs* (rather than roots) absorb water.

Improve your grade

Mineral uptake

Look at the graph.
What conclusions can be made from this data?
AO3 [4 marks]

Key: inside cells; in pond water
concentration in M (0, 0.025, 0.05, 0.075, 0.1)
chlorine, potassium, magnesium, calcium
mineral

Mineral uptake by algae growing in water

Decay

Decay

- Earthworms, maggots and woodlice are called **detritivores** because they feed on dead and decaying material (**detritus**).
- Detritivores increase the rate of **decay** by breaking up the detritus and so increasing the surface area for further microbial breakdown.
- The rate of decay can be increased by increasing the temperature, amount of oxygen and water.

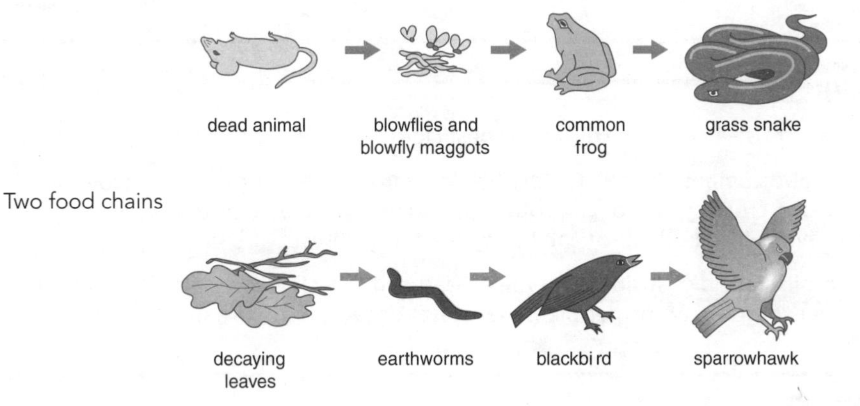

Two food chains

D–C

- Increasing the temperature to an optimum of 37 °C for **bacteria** or 25 °C for **fungi** will increase their rate of **respiration**. Higher temperatures will denature enzymes.
- By increasing the amount of oxygen, bacteria will use **aerobic respiration** to grow and reproduce faster.
- Increasing the amount of water will allow for material to be digested and absorbed more efficiently and increase growth and reproduction of bacteria and fungi.
- A **saprophyte**, for example a fungus, is an organism that feeds on dead and decaying material.
- Fungi produce enzymes to digest food outside their cells and then reabsorb the simple **soluble** substances. This type of digestion is called extracellular digestion.

B–A*

Food preservation

- Food preservation methods reduce the rate of decay.
  - In canning, foods are heated to kill bacteria and then sealed in a **vacuum** to prevent entry of oxygen and bacteria.
  - Cooling foods will slow down bacterial and fungal growth and reproduction.
  - Freezing foods will kill some bacteria and fungi and slow down their growth and reproduction.
  - Drying foods removes water so bacteria cannot feed and grow.
  - Adding salt or sugar will kill some bacteria and fungi, as the high osmotic **concentration** will remove water from them.
  - Adding vinegar will produce very acid conditions killing most bacteria and fungi.

D–C

EXAM TIP

Fungi and bacteria do not use photosynthesis to make food.

Remember!

Keeping food cool in a refrigerator only slows down the growth of bacteria. It does not stop it completely.

Improve your grade

Growing mushrooms

Lynn wants to grow mushrooms (a fungus).

Explain what conditions she should provide for optimum growth. *AO1/2* [3 marks]

Farming

Pesticides

D–C

- The use of pesticides such as **insecticides**, **fungicides** and **herbicides** has disadvantages.
  - They can enter and accumulate in food chains causing a lethal dose to **predators**.
  - They can harm other organisms living nearby which are not pests.
  - Some are persistent (take a very long time to break down and become harmless).

Remember!
Both organic and intensive farming have advantages and disadvantages.

Organic farming

D–C

- Organic farming does not use artificial **fertilisers** or pesticides.
- It uses animal manure and **compost** (instead of artificial fertilisers), **crop rotation** (to avoid build-up of soil pests), nitrogen-fixing crops as part of the rotation, and varying seed planting times to get a longer crop time and avoid certain times of the life cycle of insect pests.
- It avoids expensive fertilisers and pesticides and their disadvantages. However, the crops are smaller and the produce more expensive. Many people believe that organic crops are healthier and tastier than other crops.

Biological control

D–C

- **Biological control** uses living organisms to control pests. Examples are using ladybirds and certain wasp species to eat aphids, which damage plants.
- Biological control can avoid the disadvantages of artificial insecticides and as they use living organisms, once introduced they usually do not need replacing.
- However, many attempts at biological control have caused other problems such as the introduced **species** eating other useful species and then showing a rapid increase in their **population** so they themselves become pests and then spread into other areas or countries, e.g. the use of cane toads in Australia.
- Introducing a species into a **habitat** to kill another species can affect the food sources of other organisms in a food web, causing unexpected results.

Hydroponics and intensive farming

D–C

- **Intensive farming**, which makes use of artificial pesticides and fertilisers, is very efficient in producing large crop yields cheaply. However, intensive farming methods raise concerns about animal cruelty, as animals are kept in small areas, and about the effects of extensive use of chemicals on soil structure and other organisms.
- Plants can be grown without using soil using **hydroponics**. This system uses a regulated recycling flow of aerated water containing **minerals** and is usually done in glasshouses and polytunnels.
- Hydroponics is a type of intensive farming that is especially useful in areas of barren soil or low rainfall. Tomatoes are a common crop from hydroponics.

EXAM TIP
Hydroponics is a form of intensive farming.

B–A*

- Being a soil-free system, hydroponics has a better control over **mineral** levels and disease. Many plants can be grown in a small space. As there is no anchorage for plants when using water, artificial fertilisers are used.
- Intensive farming improves the **efficiency** of energy transfer in food chains involving humans by reducing or removing competing organisms such as animal pests and weeds. Also by keeping animals inside sheds or barns (battery farming), they use less energy to keep warm and to move, and more energy on growth (cattle) or egg production (hens).

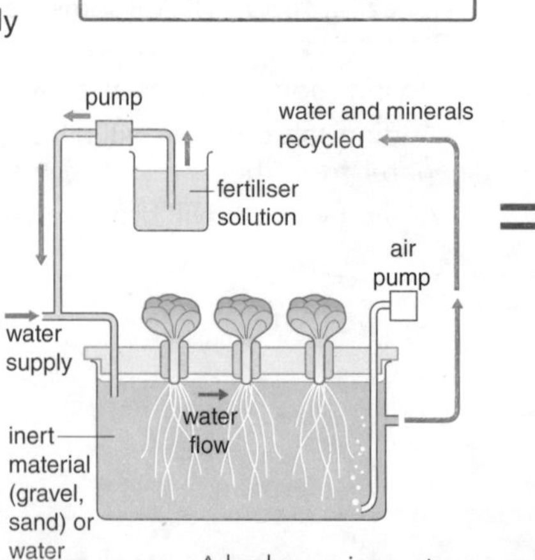

A hydroponics system

Improve your grade

Hydroponics

Look at the diagram of a hydroponics system. Explain how such a system is useful for growing plants such as lettuce in glasshouses. *AO1/2* [4 marks]

B4 Summary

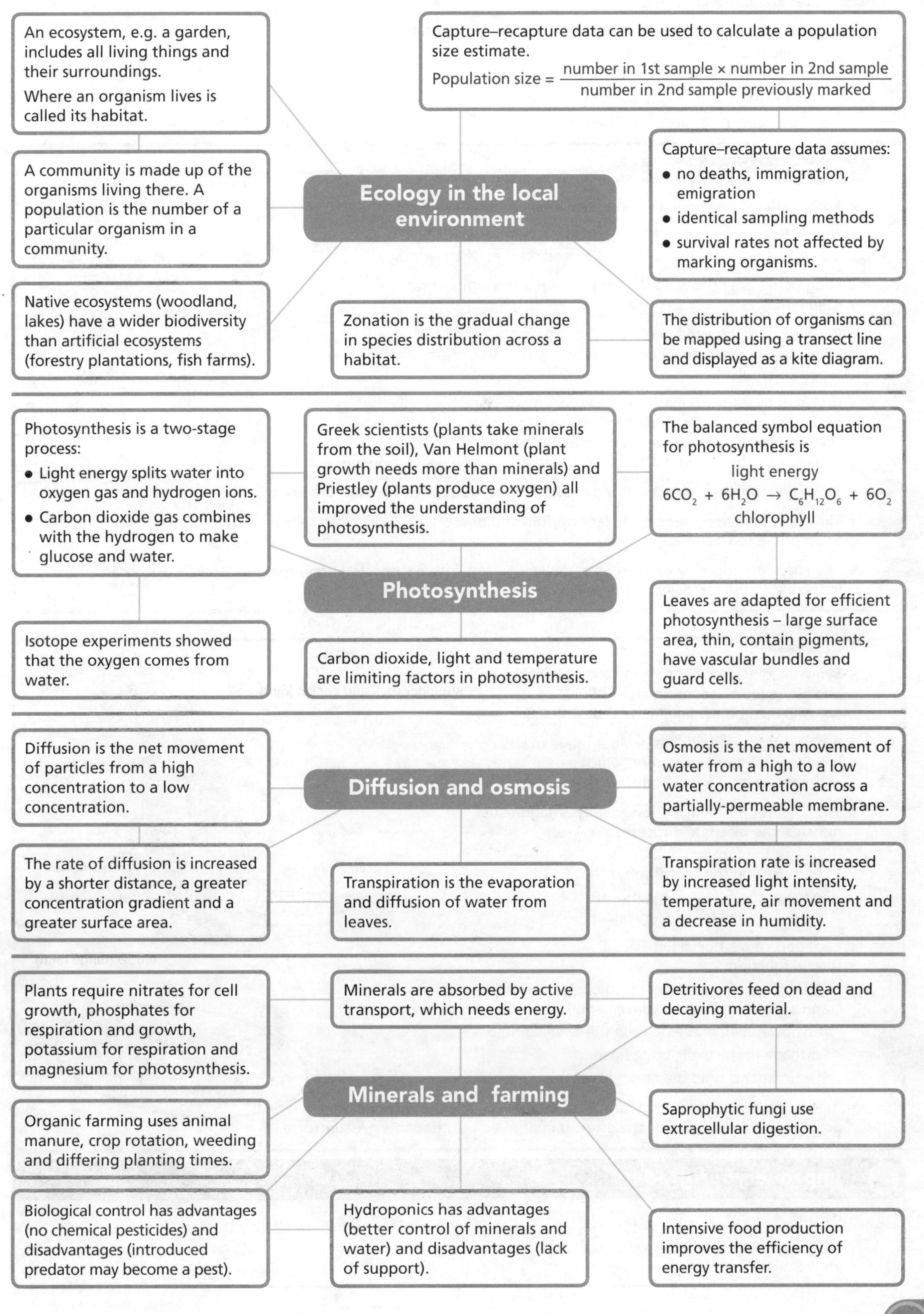
Ecology in the local environment
An ecosystem, e.g. a garden, includes all living things and their surroundings.
Where an organism lives is called its habitat.
A community is made up of the organisms living there. A population is the number of a particular organism in a community.
Native ecosystems (woodland, lakes) have a wider biodiversity than artificial ecosystems (forestry plantations, fish farms).
Capture–recapture data can be used to calculate a population size estimate.
Population size = number in 1st sample × number in 2nd sample / number in 2nd sample previously marked
Capture–recapture data assumes:
• no deaths, immigration, emigration
• identical sampling methods
• survival rates not affected by marking organisms.
Zonation is the gradual change in species distribution across a habitat.
The distribution of organisms can be mapped using a transect line and displayed as a kite diagram.
Photosynthesis
Photosynthesis is a two-stage process:
• Light energy splits water into oxygen gas and hydrogen ions.
• Carbon dioxide gas combines with the hydrogen to make glucose and water.
Greek scientists (plants take minerals from the soil), Van Helmont (plant growth needs more than minerals) and Priestley (plants produce oxygen) all improved the understanding of photosynthesis.
The balanced symbol equation for photosynthesis is
light energy
$6CO_2 + 6H_2O \rightarrow C_6H_{12}O_6 + 6O_2$
chlorophyll
Isotope experiments showed that the oxygen comes from water.
Carbon dioxide, light and temperature are limiting factors in photosynthesis.
Leaves are adapted for efficient photosynthesis – large surface area, thin, contain pigments, have vascular bundles and guard cells.
Diffusion and osmosis
Diffusion is the net movement of particles from a high concentration to a low concentration.
Osmosis is the net movement of water from a high to a low water concentration across a partially-permeable membrane.
The rate of diffusion is increased by a shorter distance, a greater concentration gradient and a greater surface area.
Transpiration is the evaporation and diffusion of water from leaves.
Transpiration rate is increased by increased light intensity, temperature, air movement and a decrease in humidity.
Minerals and farming
Plants require nitrates for cell growth, phosphates for respiration and growth, potassium for respiration and magnesium for photosynthesis.
Minerals are absorbed by active transport, which needs energy.
Detritivores feed on dead and decaying material.
Organic farming uses animal manure, crop rotation, weeding and differing planting times.
Saprophytic fungi use extracellular digestion.
Biological control has advantages (no chemical pesticides) and disadvantages (introduced predator may become a pest).
Hydroponics has advantages (better control of minerals and water) and disadvantages (lack of support).
Intensive food production improves the efficiency of energy transfer.

Skeletons

Types of skeletons

- An **internal skeleton** has many advantages over an **external skeleton**:
  - It provides an internal framework for the body.
  - It grows with the rest of the body.
  - It is flexible, due to the many joints.
  - It allows easy attachment of muscles.

Cartilage and bone

- Both **cartilage** and bone are living tissues containing living cells.
- A long bone consists of a long shaft containing bone marrow with blood vessels. At each end there is a head covered with cartilage.
- Long bones are hollow, so they are stronger and lighter than solid bones.
- Because cartilage and bone are living tissues, they can be infected by **bacteria** and **viruses**. However, they are able to grow and repair themselves.
- In very early stages, the human skeleton is made of cartilage. By the process of **ossification** (the deposition of calcium and phosphorus), the cartilage is slowly replaced by bone. If some cartilage remains between the head and shaft, the bone and the person is still growing.

cartilage
cartilage
bone
marrow

The process of ossification

- Even though bones are very strong they can easily be broken with a sharp knock.
- The bones of elderly people can lack calcium and phosphorus, which can result in **osteoporosis**, making them prone to **fractures**.
- In an accident it can be dangerous to move a person with a suspected bone fracture. Broken vertebrae in the backbone can damage the spinal cord, resulting in paralysis or death.

Joints

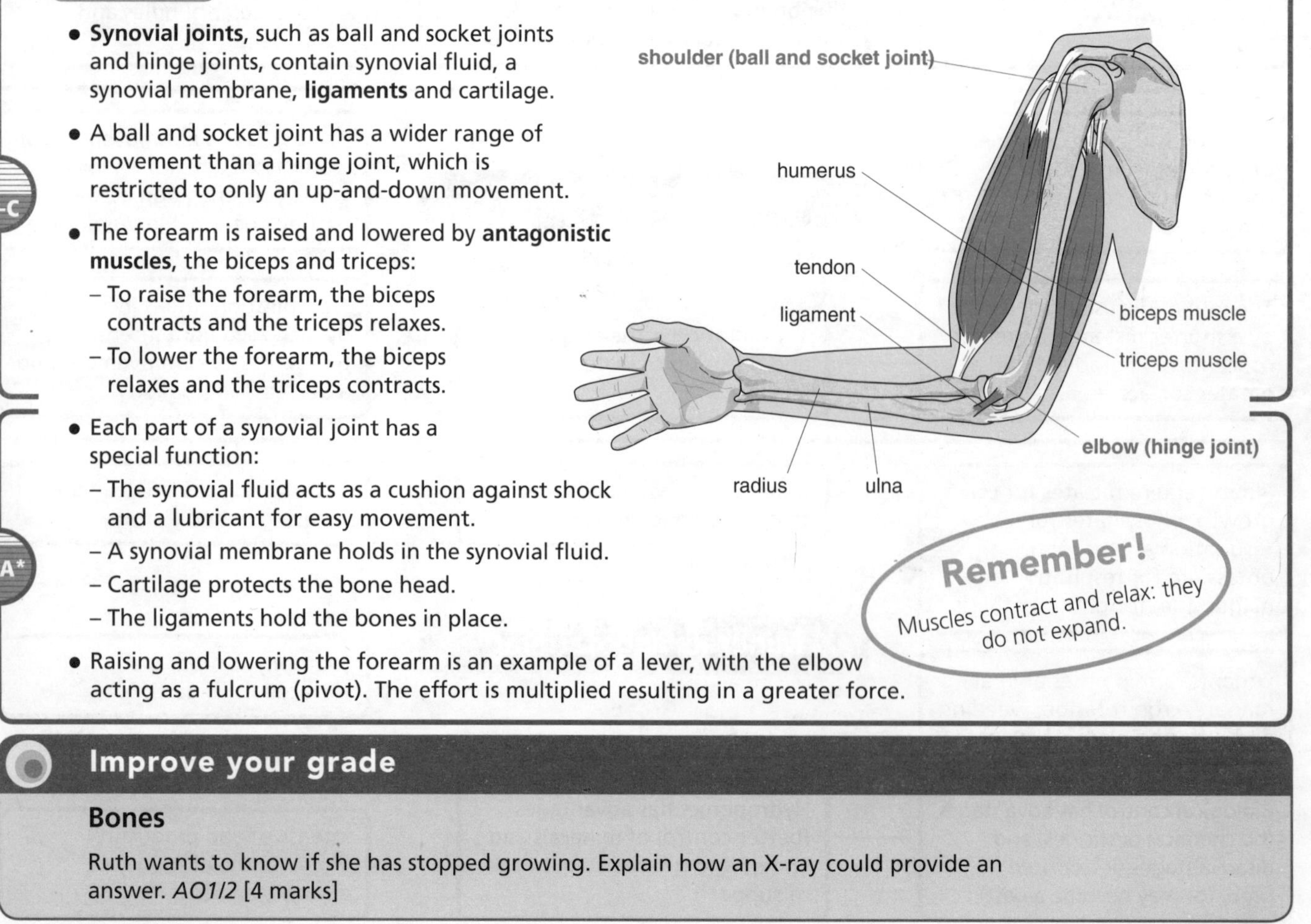

- **Synovial joints**, such as ball and socket joints and hinge joints, contain synovial fluid, a synovial membrane, **ligaments** and cartilage.
- A ball and socket joint has a wider range of movement than a hinge joint, which is restricted to only an up-and-down movement.
- The forearm is raised and lowered by **antagonistic muscles**, the biceps and triceps:
  - To raise the forearm, the biceps contracts and the triceps relaxes.
  - To lower the forearm, the biceps relaxes and the triceps contracts.
- Each part of a synovial joint has a special function:
  - The synovial fluid acts as a cushion against shock and a lubricant for easy movement.
  - A synovial membrane holds in the synovial fluid.
  - Cartilage protects the bone head.
  - The ligaments hold the bones in place.
- Raising and lowering the forearm is an example of a lever, with the elbow acting as a fulcrum (pivot). The effort is multiplied resulting in a greater force.

Remember!
Muscles contract and relax: they do not expand.

Improve your grade

Bones

Ruth wants to know if she has stopped growing. Explain how an X-ray could provide an answer. *AO1/2* [4 marks]

Circulatory systems and the cardiac cycle

Circulatory systems

- Many animals need a blood circulatory system to ensure all their cells receive enough food and oxygen and to remove waste products, such as **carbon dioxide**.
- As blood flows through arteries, veins and capillaries, the **blood pressure** decreases. Veins have valves to ensure there is no backward blood flow. High blood pressure would damage the fragile walls of the capillaries.
- A **single circulatory system** (such as in fish) has a single blood circuit of the heart, gills and body. A **double circulatory system** (such as in **mammals**) has two circuits: the heart and lungs form one (to obtain oxygen), the heart and rest of the body form the other (to deliver oxygen to body cells).

D–C

- A double circulatory system requires a four-chambered heart: two atria to receive blood (from the lungs and body) and two ventricles to distribute blood (to lungs and body). It ensures high blood pressure for efficient and fast circulation of food and oxygen. A single circulatory system needs only two chambers in the heart, one to receive and one to distribute blood.
- In the second century, Galen knew that the heart acted as a pump and the importance of the pulse. However, he thought that the liver made blood that flowed backwards and forwards. In the seventeenth century, William Harvey knew that blood circulated around the body, that the heart has four chambers and about the tiny vessels that today we know as capillaries.

B–A*

Cardiac cycle

- The cardiac cycle is the sequence of events as blood enters and leaves the heart.
- The muscles of the two atria contract together as the two ventricles relax to receive the blood through the atrio-ventricular valves, which prevent backward flow into the atria. Muscles of the two ventricles then contract together to force blood to the lungs or around the body. Semi-lunar valves prevent backward flow into the ventricles.

B–A*

Heart rate

- More muscular activity causes a greater demand for oxygen and food. Heart rate therefore increases with increasing muscular activity.
- Heart rate is increased by the presence of the hormone **adrenaline** to prepare the body for 'fight or flight'.
- Groups of cells in the heart form **pacemakers**, which control the rate of heart beat by producing a small electric current to stimulate muscle contraction. An artificial pacemaker can be placed near the heart to send an electrical signal to the heart muscle.
- An electrocardiogram (ECG) (which shows changes in electrical impulses in heart muscle) and an echocardiogram (which displays a video of the heart in action) can be used to investigate irregular heart actions.

D–C

- Two pacemakers, the sino-atrial node (SAN) and the atrio-ventricular node (AVN) generate electrical impulses to coordinate heart muscle contraction. Impulses from the SAN cause the atria to contract and stimulate the AVN. Impulses from the AVN cause the ventricles to contract.

B–A*

Improve your grade

Pressure changes in blood vessels

Look at the graph showing pressure changes in blood vessels.

(a) Explain what happens to blood pressure in arteries. *AO2* [2 marks]

(b) Describe and explain the general trend shown in the graph. *AO2* [2 marks]

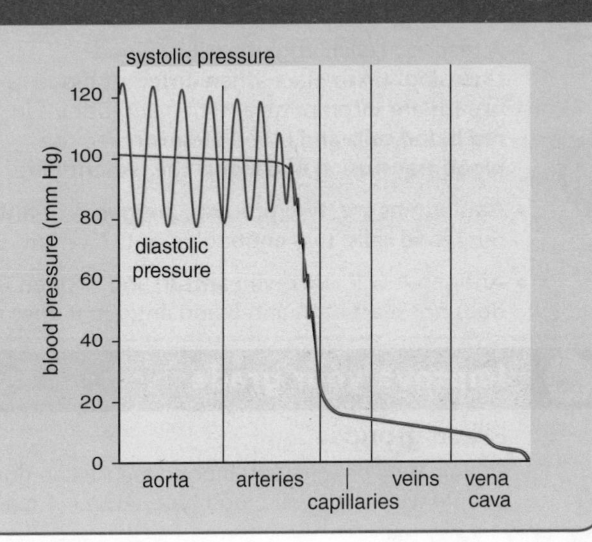

Running repairs

Heart problems

D–C

- The heart cannot work properly with a hole in the wall between the two sides of the heart. This can be corrected by 'open heart' surgery.
- A 'hole in the heart' allows blood to flow from one side of the heart to the other, so blood leaving the heart in the aorta carries less than the usual amount of oxygen, causing muscles to have less oxygen and therefore less energy.

B–A*

- A 'hole in the heart' causes the mixing of oxygenated and deoxygenated blood, resulting in the arterial blood carrying less oxygen.
- The circulation in an unborn baby is different from its circulation after it is born since the lungs do not function until it is born. An unborn baby therefore does not need a **double circulatory system**. Before birth, a hole exists between the two sides of the heart, which closes at birth.

Repairing the heart

D–C

- The heart cannot work properly with damaged or weak valves. They can be repaired or replaced by surgery. A heart with damaged or weak valves produces a lower **blood pressure** and poor circulation, as blood will leak backwards.
- A blocked **coronary artery** reduces the blood flow to the heart muscles. It can be by-passed by transplanting a blood vessel from another part of the body.
- Major heart problems can be corrected by transplanting donor hearts. Small electrical pumps (heart assist devices) can also be used to provide extra pressure to blood leaving the heart, so allowing time for damaged muscle to recover.

B–A*

- Artificial **pacemakers** and artificial valves have obvious advantages of keeping patients alive and improving their quality of life. There are also no problems of donor shortage and finding tissue matches. Surgery carries a risk, especially the major surgery of a heart transplant. Preventing rejection of a transplant also involves lifetime use of immuno-suppressive drugs.

Blood transfusions

D–C

- Blood donation involves the collection of blood from a volunteer. **Anti-coagulant** drugs such as heparin are used and the blood group and **rhesus** information is recorded.
- A **blood transfusion** puts the correct blood type into the patient's blood system, replacing blood lost after an accident or operation.
- Doctors can use drugs such as warfarin, heparin and aspirin to prevent clotting, which can block blood vessels in some medical conditions.
- People with the inherited condition haemophilia are at risk of internal bleeding from the slightest knock, as the blood does not clot.
- The process of blood clotting is called a cascade process as it involves many steps. When blood **platelets** are exposed to the air at a wound site, it triggers a complex sequence of chemical reactions, eventually leading to the formation of a meshwork of fibrin fibres (clot).

B–A*

- A reaction called **agglutination** (blood clumping) takes place when different blood groups are incompatible. When agglutinins in **red blood cells** and blood plasma react, the blood transfusion endangers the patient's life.

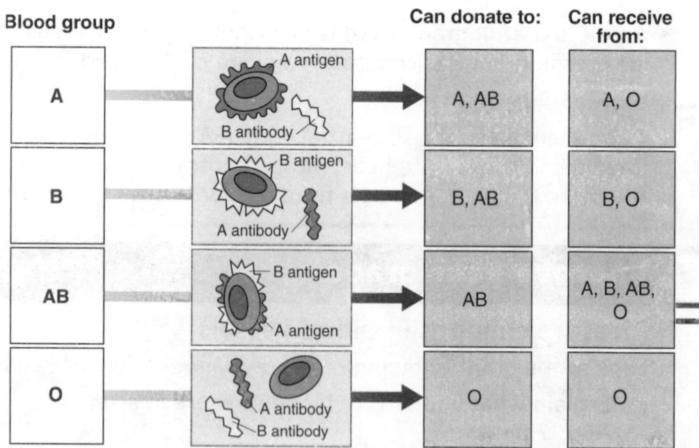

Blood group agglutinins

- Agglutinins are: two proteins, **antigen** A or antigen B on the surface of red blood cells; two **antibodies**, anti-A or anti-B in blood **plasma**.
- Antigen A will react with anti-A, and antigen B will react with anti-B, causing agglutination. Antigen A does not react with anti-B and antigen B does not react with anti-A.

Improve your grade

Blood groups

Explain why people with blood type AB can donate blood only to people with the same blood type. *AO2/3* [4 marks]

Remember!
A 'hole in the heart' is normal in an unborn baby.

Respiratory systems

Types of respiratory system

D–C

- Because of their respiratory systems, amphibians are restricted to moist **habitats**. Fish gills only work in water.

B–A*

- Amphibians have simple lungs, but use their moist, permeable skin to obtain oxygen. The permeable skin makes them susceptible to excessive water loss, which can result in death. Fish gills obtain oxygen from water being forced over filaments, so do not work in air.

Gaseous exchange

D–C

- In humans, breathing in depends on the contraction of the intercostal muscles (which connect the ribs) and the muscles in the diaphragm. The ribs are moved up and out and the diaphragm is moved downwards, causing the chest volume to increase and the pressure to decrease. The higher outside pressure causes air to enter the lungs.
- To breathe out, the intercostal muscles and diaphragm relax, causing the ribs to move down and inwards and the diaphragm to curve upwards. The chest volume therefore decreases, which increases the pressure, forcing the air out of the lungs.
- The total lung capacity consists of:
  - tidal air, which is the amount of air normally breathed in and out while at rest
  - vital capacity, which is the maximum amount of air which can be exchanged
  - residual air, which is the amount of air which cannot be forced out of the lungs.
- Exchange of gases takes place by **diffusion** between the alveoli (bulges of air sacs) and the air in the air sacs. Diffusion occurs because the oxygen concentration in the air is higher than in the deoxygenated blood capillaries around the alveoli.

B–A*

- The exchange surfaces are adapted for efficient gas exchange. They have a large surface area and a good blood supply. They are permeable, moist and only one cell thick.
- Readings from a spirometer are a measure of different lung capacities and the rate of air flow. They can be used to help diagnose lung diseases.

Respiratory diseases

D–C

- Asbestosis is an industrial disease resulting from breathing in asbestos fibres. The fibres cause inflammation and scarring of lung tissue, reducing **gaseous exchange**. In the inherited condition cystic fibrosis, too much mucus is produced in the bronchioles causing breathing difficulties.
- The risk of lung cancer is greatly increased by lifestyle factors such as smoking. Lung cells grow rapidly, reducing the surface area available for gaseous exchange.
- **Asthma** symptoms are wheezing, a tight chest and difficulty in breathing. Inhalers can be used to relieve these symptoms.
- The respiratory system in fish allows for a through flow of water. In humans, air must go in and out of the same structures. This means that chemical particles such as tars (from cigarettes) and asbestos fibres can become trapped in air sacs.

B–A*

- During an asthma attack, the lining of airways becomes inflamed, mucus and fluid build up in airways and the muscles around the bronchioles contract, narrowing the airways.

Improve your grade

Spirometer recording

Look at the diagram showing a spirometer recording of a healthy adult.

(a) Work out this person's (i) total lung capacity and (ii) vital capacity. *AO2* [2 marks]

(b) Spirometer readings are not usually taken during an asthma attack. If they were, suggest what changes in recordings would be expected and explain why they occur. *AO3* [2 marks]

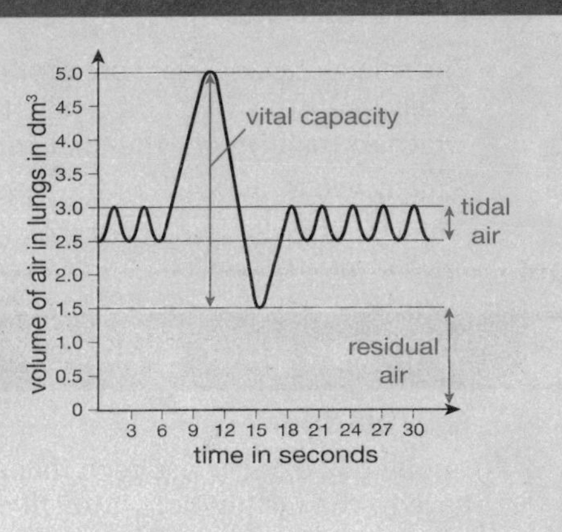

Digestion

Physical digestion

D–C

- **Physical digestion** (breaking food into smaller pieces) is important because:
  - it allows the food to pass more easily through the **digestive system**
  - it prepares the food for **chemical digestion** by giving it a larger surface area.

Chemical digestion

D–C

- In chemical digestion, **carbohydrates**, fats and proteins are broken down by specific **enzymes**, in three places:
  - in the mouth, where carbohydrase breaks down starch to sugar
  - in the stomach, where **protease** breaks down protein to amino acids
  - in the small intestine, where **lipase** breaks down fat into fatty acids and **glycerol**.

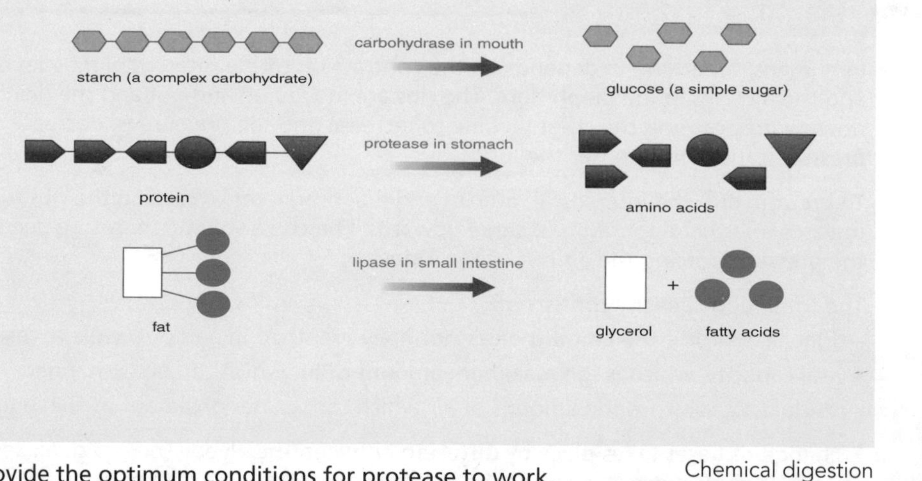

Chemical digestion

- The strong stomach acids provide the optimum conditions for protease to work.

B–A*

- The chemical breakdown of starch involves two steps:
  - breakdown of starch (many linked units) to maltose (two linked units)
  - breakdown of maltose to glucose (one unit).
- Protease enzymes, such as pepsin in the stomach, require a low **pH** (acidic) while other protease enzymes in the mouth and small intestine require a high pH (alkaline). Those in the mouth require a slightly acid/neutral pH.
- The gall bladder stores bile. Bile is released into the small intestine to emulsify fats, increasing their surface area for efficient digestion.

Remember!
Chemical digestion requires enzymes of different pH ranges.

Food absorption

D–C

- Food **molecules** need to be able to pass through the walls of the small intestine and dissolve in the blood or lymph. This means they have to be small and soluble.
- The digested carbohydrates and protein molecules (glucose and amino acids) are soluble. They diffuse through the walls of the small intestine and into the blood.
- The digested fat molecules (fatty acids) are not soluble in water or plasma, so would block up blood vessels. They diffuse through the walls of the small intestine and into the lymph.

digested food
lacteal
villus
microvilli
blood capillaries
section through small intestine
blood system (for glucose, amino acids)
lymphatic system (for fatty acids)

Food absorption

B–A*

- The small intestine is adapted for efficient absorption of food by having an extensive system of blood capillaries and an extensive lymphatic system of lacteals, which contain lymph. The small intestine also has a large surface area, created by:
  - many **villi** in the walls of the small intestine
  - many **microvilli** (projections) from the walls of the villi.

Improve your grade

Gall bladder

Jason's gall bladder has been removed in an operation. Explain why he may have to change his diet. *AO1/2* [3 marks]

Waste disposal

Kidney structure and function

- Excess and unwanted amino acids are broken down in the liver, forming **urea**, which is taken in the blood to the kidneys.
- Blood containing waste such as urea enters each kidney by the renal artery. Blood without waste leaves by the renal vein. Each kidney has an outer cortex and an inner medulla. Waste removed from the kidney leaves through the ureters as urine.
- The blood flows through the kidneys under high pressure so filtration to remove wastes also takes place under high pressure. Useful materials such as water, glucose and salt are reabsorbed back into the blood.

Remember! Urine contains urea, as well as water and salt.

D–C

- Each kidney has millions of microscopic kidney tubules (nephrons) where filtration takes place to form urine.
- Each nephron has:
  - a network of capillaries (the **glomerulus**) surrounded by a capsule: this forms a filtration unit
  - a region where some materials such as glucose are selectively reabsorbed
  - a region where reabsorption of some salt and water takes place (the amount depends on body demands).
- A dialysis machine is used when someone has kidney failure. The machine has many tubes containing blood, surrounded by a liquid. The machine acts as an artificial kidney and removes urea from the blood. As urea **molecules** are small, they diffuse through the membrane. A dialysis machine also uses different sizes of tubes, so it slightly increases pressure during **diffusion**.
- The dialysis fluid contains sodium salts, so it is the same or slightly lower than the desired blood concentration. This maintains the sodium levels in the blood.

B–A*

Regulating urine concentration

- After drinking a large quantity of water, the quantity of urine produced increases and the urine concentration decreases.
- During strenuous exercise or in hot conditions the body produces more sweat to cool down, so the quantity of urine produced decreases and the urine concentration increases.

D–C

- The pituitary gland produces the **anti-diuretic hormone** (ADH) which controls the concentration of urea by:
  - increasing the permeability of kidney tubules so that more water is reabsorbed
  - using a negative feedback mechanism to control ADH production.

B–A*

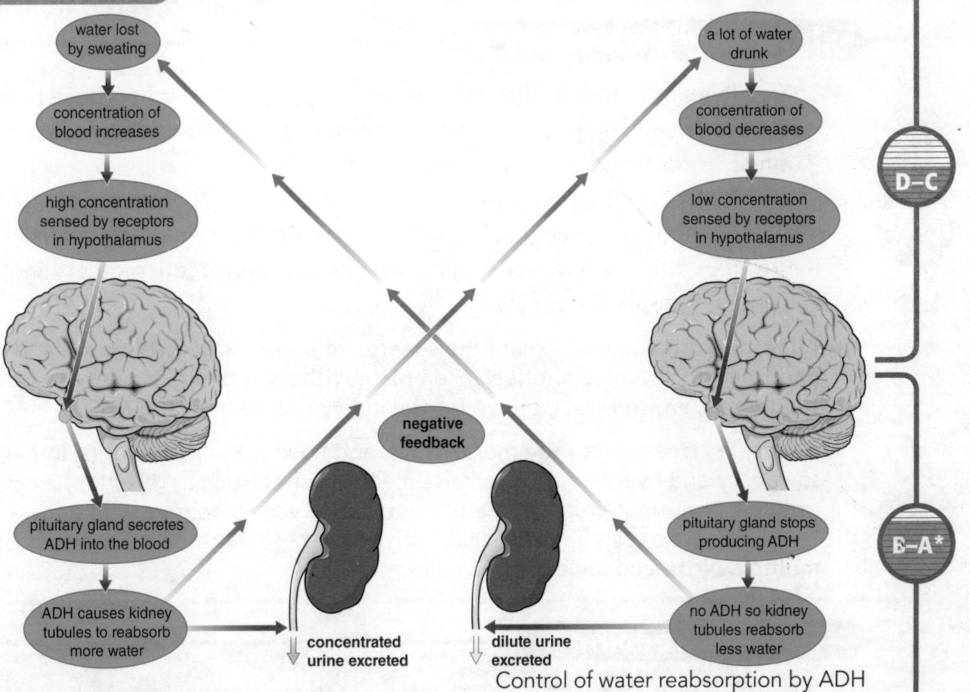

Control of water reabsorption by ADH

Carbon dioxide concentration

- **Carbon dioxide** at high concentrations is toxic and must be removed from the body.

D–C

- The body is more sensitive to the level of carbon dioxide than to that of oxygen. An increase in carbon dioxide in the blood is detected by receptors in the carotid artery. Nerve impulses inform the brain, which causes the rate of breathing to increase to remove more carbon dioxide via the lungs.

B–A*

Improve your grade

Carbon dioxide concentration

Sonya uses her exercise machine. Her increased rate of breathing will get more oxygen into her body. Explain another reason why her breathing rate will increase.
AO1/2 [4 marks]

Life goes on

Role of sex hormones in the menstrual cycle

D–C

- Four sex hormones control the female **menstrual cycle**:
  - **FSH (follicle stimulating hormone)** stimulates an egg to develop in an ovary.
  - **LH (luteinising hormone)** controls ovulation (egg release).
  - **Progesterone** maintains the uterus wall.
  - **Oestrogen** repairs the uterus wall.
- FSH and LH are released from the pituitary gland in the brain.

B–A*

- Negative feedback mechanisms (which restore the situation after change) control the levels of the sex hormones in the menstrual cycle. The cycle is triggered by the receptors in the hypothalamus.
- If **fertilisation** does not occur, the levels of oestrogen and progesterone decrease.
- When oestrogen and progesterone levels are low, menstruation occurs.
- A message is sent to the hypothalamus that hormone levels are again low. This starts the cycle to begin again.
- If an egg is fertilised, the levels of progesterone remain high and no FSH is produced, so no more eggs develop and the uterus lining does not break down.

Fertility in humans

D–C

- Fertility in humans can be controlled by the use of artificial sex hormones by controlling egg release and implantation. The contraceptive pill prevents ovulation and fertility drugs help to ensure ovulation.

B–A*

- Artificial sex hormones prevent ovulation by making the body think it is pregnant and this inhibits FSH release. Eggs in the ovary are therefore not stimulated to develop.

Infertility treatments

D–C

- There are many methods of treating infertility (the inability to produce babies):
  - artificial insemination, where sperm are placed into the vagina by syringe
  - using FSH to stimulate egg development
  - IVF ('in vitro' fertilisation),where an egg is fertilised by sperm outside the body
  - egg donation, where an egg is donated from another female, then fertilised and placed inside the uterus
  - surrogacy, where a fertilised egg is placed inside a surrogate mother (another female)
  - an ovary transplant from another female.
- All fertility treatments increase the chances of a successful fertilisation and pregnancy. This is very important for couples who feel incomplete without a family. However, not all people agree with such treatments, which are expensive for the individuals and for the NHS.

B–A*

- All fertility treatments raise moral issues, and have risks and benefits. In particular, egg donation, surrogacy and ovary transplants raise medical issues (such as rejection) as well as moral ones (for example about paternity). Some treatments are very expensive with a low rate of success. A single IVF treatment cycle costs about £6 000, with an average success rate of about 25 per cent and the risk of multiple births and lower birth weight.

Foetal screening

D–C

- A developing foetus can be checked to see if there are any abnormalities (e.g. in its growth or genetic makeup). Checking can be done by:
  - **amniocentesis** (extracting and testing cells in the amniotic fluid)
  - chromosomal analysis (using a blood test to test cells for any **chromosome** abnormalities).
- Using techniques like this raise ethical issues: whether it is right to interfere with a natural process and whether an unborn foetus has the right to life. The techniques also carry a small risk of causing the expulsion of the foetus.

Improve your grade

IVF treatment

Describe what is involved in IVF and explain how it can solve some infertility problems. *AO1/2* [4 marks]

Remember!
FSH is a female sex hormone; IVF is an infertility treatment.

Growth and repair

Growth rate

- A balanced diet (containing calcium, phosphorus, vitamin D and proteins) and regular exercise can increase growth.
- Extremes of height are usually caused by **hormone** imbalance or by **genes**.
- Different parts of a foetus and a baby grow at different rates, e.g. the head grows and develops earlier than the rest of the body.
- A baby's length, mass and head size are regularly monitored to give an early warning of any growth difference from normal, possibly due to malnourishment or hormone imbalance. Average growth charts are used for comparison.

D–C

- The human **growth hormone** is produced by the pituitary gland and it stimulates general growth, especially in long bones.

B–A*

Life expectancy

- Human life expectancy has increased in recent times due to:
  - fewer deaths from industrial diseases
  - better housing, so there are fewer cases of diseases such as tuberculosis
  - a healthier diet and lifestyle
  - advances in modern medicine, such as **antibiotics** and transplants.

Remember!
An ageing population raises ethical questions.

D–C

- More people living longer has many personal and national consequences, such as a longer retirement to enjoy but a bigger burden on pension funds and health services.

B–A*

Organ donation

- The supply of donated organs is limited by a shortage of donors and also by restrictions of use due to the necessity of tissue matches as well as those of size and age.
- These problems can be avoided by using mechanical replacements. However, these have other problems such as the dependence on a power supply, the properties of materials used, their large size and body reactions to 'foreign' materials.
- A living person can donate blood and **bone marrow** (as the body replaces them) and a kidney (as we can live with only one). Transplants require a suitable tissue match.
- Organ donations from a dead donor must meet certain criteria, such as approval from the donor or relatives, and the requirement that the donor is 'brain dead'.
- Organ donation, especially from dead donors, raises ethical issues to do with human rights, the acceptance of surgery on a dead body and the fact that a person's death has been necessary to supply a donor organ.

D–C

- Transplants are at risk of being rejected by the recipient's body and so need life-long immuno-suppressive drug treatment, which can lead to the body not being able to protect itself from **microorganisms**.
- People over 18 years of age can ask to be put on the donor register so their organs can be used after death. There are long waiting lists for organs and deaths due to shortages. Some countries have an 'opt out' system, which assumes that organs can be donated without asking permission. Some people object to this system, saying it is against human rights.
- Trends in transplant and survival rates can be shown by interpreting data.

B–A*

Improve your grade

Transplant survival rates

Look at the graph showing survival rates from transplants in the UK in 2010. Suggest why different organ transplants have different survival rates. *AO2* [3 marks]

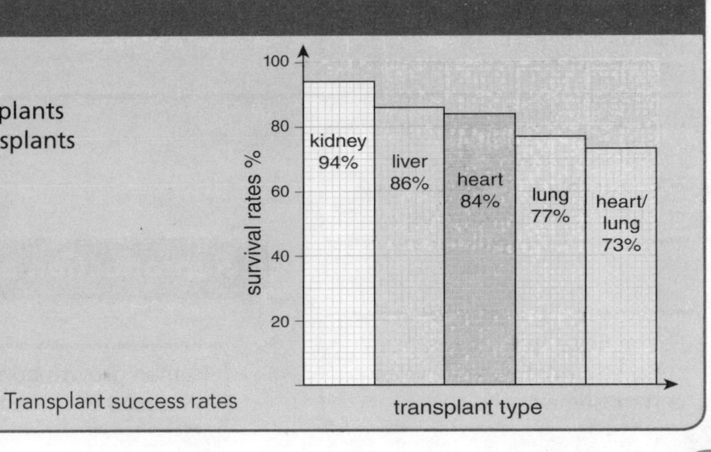

Transplant success rates

B5 Summary

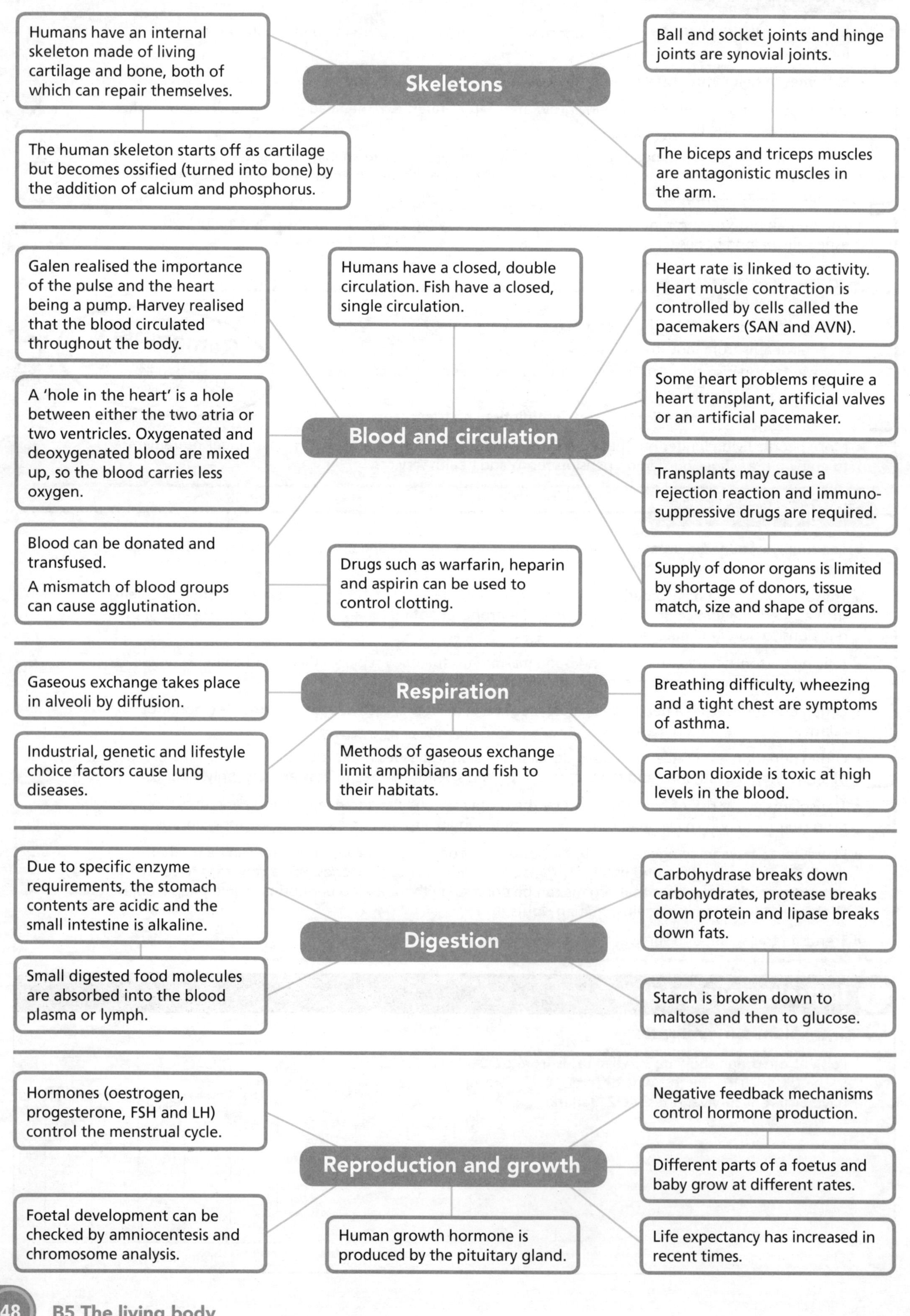

Skeletons
Humans have an internal skeleton made of living cartilage and bone, both of which can repair themselves.
The human skeleton starts off as cartilage but becomes ossified (turned into bone) by the addition of calcium and phosphorus.
Ball and socket joints and hinge joints are synovial joints.
The biceps and triceps muscles are antagonistic muscles in the arm.
Blood and circulation
Galen realised the importance of the pulse and the heart being a pump. Harvey realised that the blood circulated throughout the body.
Humans have a closed, double circulation. Fish have a closed, single circulation.
Heart rate is linked to activity. Heart muscle contraction is controlled by cells called the pacemakers (SAN and AVN).
A 'hole in the heart' is a hole between either the two atria or two ventricles. Oxygenated and deoxygenated blood are mixed up, so the blood carries less oxygen.
Some heart problems require a heart transplant, artificial valves or an artificial pacemaker.
Transplants may cause a rejection reaction and immuno-suppressive drugs are required.
Blood can be donated and transfused.
A mismatch of blood groups can cause agglutination.
Drugs such as warfarin, heparin and aspirin can be used to control clotting.
Supply of donor organs is limited by shortage of donors, tissue match, size and shape of organs.
Respiration
Gaseous exchange takes place in alveoli by diffusion.
Breathing difficulty, wheezing and a tight chest are symptoms of asthma.
Industrial, genetic and lifestyle choice factors cause lung diseases.
Methods of gaseous exchange limit amphibians and fish to their habitats.
Carbon dioxide is toxic at high levels in the blood.
Digestion
Due to specific enzyme requirements, the stomach contents are acidic and the small intestine is alkaline.
Carbohydrase breaks down carbohydrates, protease breaks down protein and lipase breaks down fats.
Small digested food molecules are absorbed into the blood plasma or lymph.
Starch is broken down to maltose and then to glucose.
Reproduction and growth
Hormones (oestrogen, progesterone, FSH and LH) control the menstrual cycle.
Negative feedback mechanisms control hormone production.
Different parts of a foetus and baby grow at different rates.
Foetal development can be checked by amniocentesis and chromosome analysis.
Human growth hormone is produced by the pituitary gland.
Life expectancy has increased in recent times.

Understanding microbes

Bacteria

- Bacterial cells have certain features that allow them to survive. These include:
  - a flagellum for movement
  - a cell wall to maintain shape and to stop it from bursting
  - **DNA** to control the cell's activities and replication of the cell.
- Bacterial cells can be different shapes such as spherical, rod shaped, spiral or curved rods.

A bacterial cell

- **Bacteria** reproduce by splitting into two in a type of **asexual reproduction** called **binary fission**.
- It is possible to get bacteria to reproduce on an agar plate but all equipment must be sterilised first to prevent contamination by other **microbes**. This is called an **aseptic technique**.
- In terms of numbers, bacteria are very successful for several reasons:
  - They can survive on an enormous range of different energy sources.
  - They can live in a very wide range of habitats.
  - Some bacteria live by taking in organic nutrients but others can make their own food.
- Bacteria can reproduce very quickly. This means that they can very rapidly spoil food or cause disease. This means that they must be handled carefully to avoid contamination of people, animals or food.

Remember!
Bacteria do not increase by growing. Numbers increase when bacteria reproduce.

Yeast

- **Yeast** is a single-celled **fungus** that is grown for many functions. Its growth rate can be altered by:
  - changing food availability
  - changing temperature
  - changing pH
  - removing waste products.
- The growth rate of yeast doubles for every 10 °C rise in temperature until the **optimum** is reached.

Viruses

- **Viruses** are not living cells but very small structures made of a protein coat surrounding a strand of genetic material.
- Viruses can only reproduce under certain conditions:
  - They only reproduce in other living cells.
  - They only attack specific cells, which may be plant, bacterial or animal cells.
- When a virus reproduces it will:
  - attach itself to a specific host cell
  - inject its genetic material into the cell
  - use the cell to make the components of new viruses
  - cause the host cell to split open and die to release the viruses.

Improve your grade

Viruses

Although they do not feed, viruses are usually described as parasites. Explain why this is.
AO2 [4 marks]

Harmful microorganisms

Disease transmission

- Disease-causing **microorganisms** can be passed on in a number of different ways. Understanding how they are passed on is very important in preventing the spread of diseases.
  - Some **microbes**, such as *Salmonella*, are spread in food. They can be prevented from spreading by correct food hygiene.
  - Some, such as *Vibrio cholera*, may be spread in water. They can be prevented from spreading by correct water treatment.
  - Other microbes need direct contact and they can be prevented from spreading by barrier methods.
  - Many microbes, such as those that cause influenza ('flu), are spread in airborne droplets. They can be stopped from spreading by correct use of paper tissues and isolation of patients.

D–C

- There are four stages in an infectious disease:
  - The microbe enters the body.
  - It reproduces many times without causing symptoms. This is the **incubation period**.
  - The microbes cause the production of many toxins.
  - The toxins cause symptoms, such as fever.
- We are always hearing about diseases occurring in areas that have experienced natural disasters. This can be for a number of reasons:
  - Damage to sewage systems may lead to water supplies being contaminated.
  - Damage to electrical supplies may stop refrigerators working, so food **decays**.
  - Large numbers of people moving to other areas means the facilities are not able to cope.
  - Hospitals may be damaged, or there may be a shortage of medical staff.

B–A*

- Doctors and health professionals collect data on the incidence of various diseases, such as influenza, food poisoning and cholera, to try and see patterns and make predictions.

Pioneers and the treatment of disease

D–C

- Many scientists have made important discoveries that have helped to prevent microbes causing disease. Three of the most important of these are:
  - Louis Pasteur, who helped to prove the germ theory of disease by realising that microbes from the air could make food go bad
  - Joseph Lister, who invented the first **antiseptic**, using carbolic acid to prevent wounds becoming infected
  - Sir Alexander Fleming, who discovered the first **antibiotic**, penicillin, which is produced from *Penicillum*, a **fungus**.

EXAM TIP

You may be asked to interpret data on diseases such as influenza, food poisoning and cholera. Make sure you can describe trends and suggest reasons for them.

- Since their discovery, antiseptics and antibiotics have been widely used to control disease.
  - Antiseptics are used on the outside of the body to kill microbes and prevent their entry.
  - Antibiotics tend to be used inside the body to kill microbes once they have entered.
  - Antiseptics work on most microbes but antibiotics have no affect on **viruses**.
- Problems are occurring now because some strains of **bacteria** are developing resistance to antibiotics. This resistance appears in bacterium by a **mutation**. Because the bacteria can then survive and reproduce, the resistance is spread by **natural selection**.

B–A*

- To try and prevent antibiotic resistance spreading, doctors take various steps:
  - They only prescribe antibiotics when really necessary.
  - They advise patients to always finish the dose so partially resistant bacteria are killed.

Improve your grade

Barbeque time

In the summer, when more food is cooked on barbeques, more people get food poisoning, (e.g. from *Salmonella*). Explain why this is and what can be done to prevent this problem.
AO2 [3 marks]

Useful microorganisms

Yoghurt making

D–C

- The process of making yoghurt uses **bacteria** and includes a number of steps:
  - First, all the equipment is sterilised.
  - Then milk is **pasteurised** by heating it to about 78 °C.
  - When the milk is cooled down it is incubated with a culture of bacteria.
  - This is followed by sampling and then adding flavours, colours and packaging.

Remember!
It is important to make sure that the milk is cooled down before adding the yeast, or the yeast will be killed.

B–A*

- The type of bacterium that is added to the milk is *Lactobacillus*. This causes the breakdown of lactose in milk to lactic acid, which makes the yoghurt taste acidic.

Fermentation

D–C

- The process of **fermentation** in **yeast** involves **anaerobic respiration**. The word equation for this reaction is:

  glucose (sugar) → ethanol (alcohol) + carbon dioxide.

- Using this reaction, yeast can be used in brewing beer or wine:
  - First, sugar is extracted by crushing grapes (wine) or from barley grains (beer).
  - Then yeast is added.
  - It is kept warm to allow it to ferment. Air and other microorganisms are kept out.
  - The wine or beer is allowed to clarify (clear). The clear liquid is then drawn off (removed from the yeast sediment).
  - The wine or beer may then be pasteurised and put into casks or bottles.
- The concentration of **alcohol** made by fermentation is limited, so to make drinks like whisky and brandy the process of **distillation** is used. This increases the alcohol concentration but is only allowed in licensed premises.

EXAM TIP
You may be asked to interpret data on yeast fermentation. Make sure you can explain how any changes in conditions could affect the yeast.

B–A*

- The balanced chemical equation for fermentation is:

  $C_6H_{12}O_6 \rightarrow 2C_2H_5OH + 2CO_2$

- When yeast is used in brewing it soon uses up all the oxygen in the container by respiring **aerobically**. This allows the number of cells to increase rapidly. Then conditions are kept anaerobic so that alcohol is made.
- Yeast breaks down sugar at different rates in different conditions, such as temperature, and the presence or absence of oxygen.
- The process of pasteurisation is used in brewing to kill harmful microbes. The liquid is kept at an elevated temperature for a predetermined time. The temperature and time depends on the drink that is being brewed.
- The alcohol concentration produced by brewing is limited. This is because high concentrations of alcohol kill yeast cells, although some strains of yeast are more resistant to alcohol than others.

Improve your grade

Yoghurt making

Brenda decides to make her own yoghurt by adding a small amount of yoghurt that she bought to some milk. Her friend Sarah says that this will not work because the bought yoghurt had been pasteurised before it was packaged.

Explain why Sarah was correct. *AO2* [3 marks]

Biofuels

Why use biofuels?

- There are several different **biofuels**. They all use the energy that is trapped in **biomass**.
  - Fast-growing trees are grown and then the wood is burnt.
  - Biomass such as sugar or waste material is **fermented** using **bacteria** or **yeast** and the product is used as fuel.
- Biofuels have become more popular for several reasons:
  - They are alternative sources to **fossil fuels**, which are running out.
  - Their waste makes no net increase in greenhouse gas levels.
  - They do not release **particulates** when they are burnt.
- There is no net increase in **greenhouse gas** levels if biofuels are burnt at the same rate as the biomass is being produced. This is sometimes called 'carbon neutral'.
- However, problems can occur if areas of land are cleared of other plants in order to grow crops for biofuels:
  - This may mean that the fuel is not carbon neutral because the other plants cannot now remove **carbon dioxide**.
  - Important **habitats** may be lost and **species** may become **extinct**.

Biogas

- **Biogas** is a fuel that contains:
  - mainly methane
  - some carbon dioxide
  - very small amounts of hydrogen, nitrogen and hydrogen sulfide.
- Biogas can be produced on a large scale in a digester. This uses a continuous flow method as organic wastes are constantly added and the gas and remaining solids constantly removed.

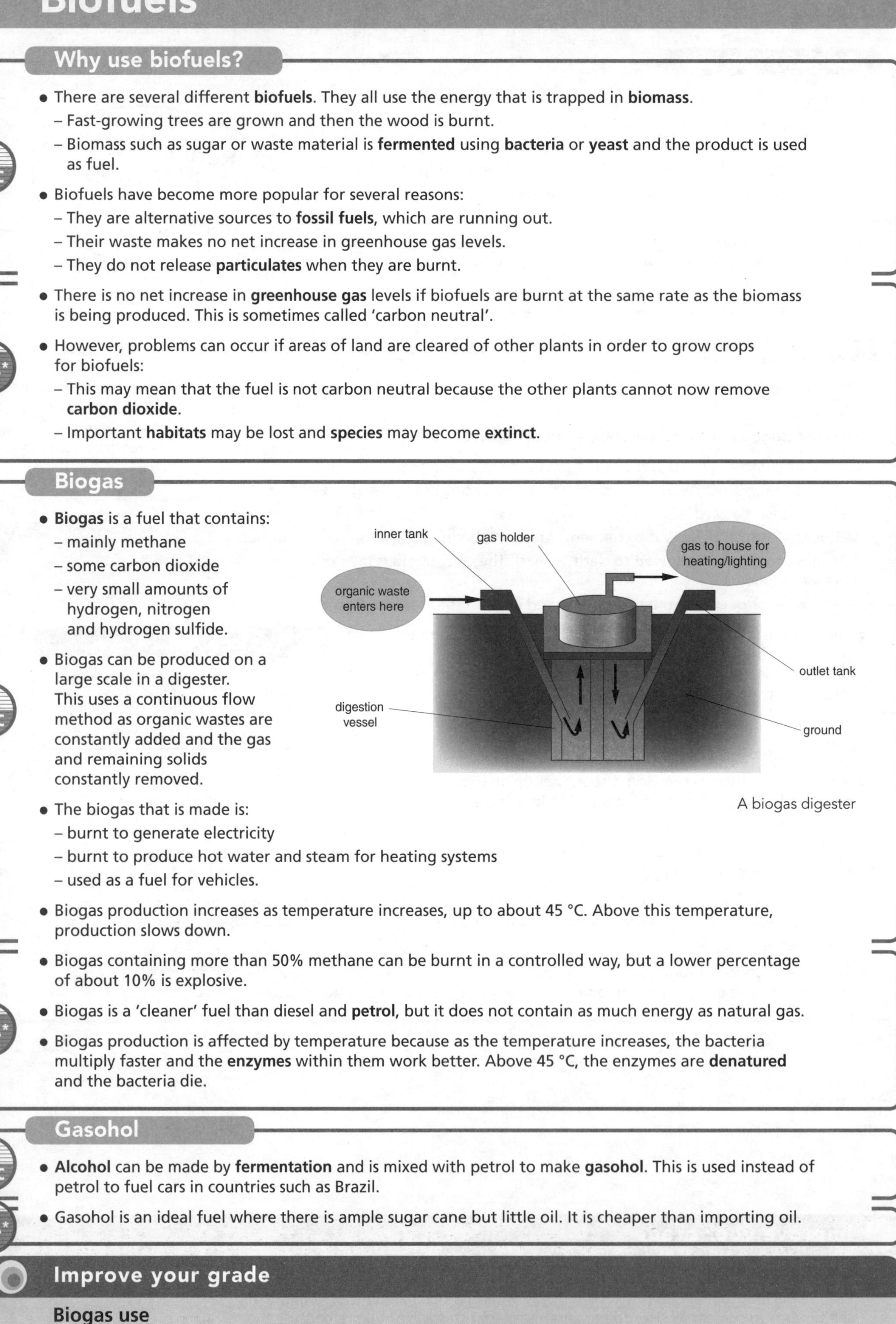

A biogas digester

- The biogas that is made is:
  - burnt to generate electricity
  - burnt to produce hot water and steam for heating systems
  - used as a fuel for vehicles.
- Biogas production increases as temperature increases, up to about 45 °C. Above this temperature, production slows down.
- Biogas containing more than 50% methane can be burnt in a controlled way, but a lower percentage of about 10% is explosive.
- Biogas is a 'cleaner' fuel than diesel and **petrol**, but it does not contain as much energy as natural gas.
- Biogas production is affected by temperature because as the temperature increases, the bacteria multiply faster and the **enzymes** within them work better. Above 45 °C, the enzymes are **denatured** and the bacteria die.

Gasohol

- **Alcohol** can be made by **fermentation** and is mixed with petrol to make **gasohol**. This is used instead of petrol to fuel cars in countries such as Brazil.
- Gasohol is an ideal fuel where there is ample sugar cane but little oil. It is cheaper than importing oil.

Improve your grade

Biogas use

Discuss the advantages and disadvantages of biogas compared to diesel and natural gas.
AO1 [4 marks]

Life in soil

The components of soil

- Soil contains **mineral** particles of different sizes. In a sandy soil the particles are smaller than in a clay soil.
- Loam is a soil that contains:
  - a mixture of clay and sand
  - a large amount of partly **decomposed** animal and plant waste called **humus.**
- Simple experiments can be performed on different soil samples to compare the contents:
  - Humus content can be found by burning off the humus using a Bunsen burner.
  - Air content can be found by seeing how much water is needed to fill the air spaces.
  - Water content can be found by slowly heating the soil to **evaporate** the water.

D–C

- If a soil has larger particles, then the air content and permeability is usually higher.
- If a soil has larger amounts of humus it will often hold more water and air.

B–A*

Living in soil

- Many organisms live in soil and depend on a supply of oxygen for **respiration** and water for chemical reactions. These organisms form many food webs.
- Humus in the soil is important to living organisms because it will:
  - decompose to release minerals
  - increase the air content of the soil.
- Earthworms are also important to soil structure and fertility because they:
  - bury organic material for decomposition by **bacteria** and **fungi**
  - **aerate** and drain the soil
  - mix up soil layers
  - neutralise acid soil.

D–C

EXAM TIP

Be prepared to look at diagrams of soil food webs and work out the relationships between the organisms.

- The aeration and draining produced by earthworms will allow organisms to respire **aerobically**.
- Neutralising acid soils is important because some plants will not grow if the **pH** is too low and mixing up soil layers is important so that dead material is decomposed.
- Many of these important functions of earthworms were first understood by Charles Darwin.

B–A*

Improve your grade

Soil analysis

A student analyses two different soils. The table shows his results.

Soil	pH	Air content	Humus content
A	5.8	13.2	5.4
B	7.2	18.5	12.8

Which soil is likely to contain more earthworms? Justify your answer.
AO3 [2 marks], *AO2* [2 marks]

Microscopic life in water

Advantages and disadvantages of living in water

D–C

- Living in water has a number of advantages.
  - There is no risk of water shortage and dehydration.
  - The temperature of the water varies less than air temperature.
  - Water helps provide support.
  - Waste products are easily disposed of into the water.
- However, it also has some disadvantages.
  - The water content of the body can vary and needs to be controlled.
  - Water is denser than air and so resists movement.

B–A*

- If the water is freshwater, then organisms can take up too much water by **osmosis**. In salt water, too much water may be lost to the surroundings by osmosis.
- Organisms such as amoeba have a **contractile vacuole** that can store any excess water. The vacuole can then fuse with the cell membrane and empty the water to the outside.

Variations in the numbers of organisms

D–C

- The numbers of **phytoplankton** (tiny aquatic plants) and zooplankton (tiny aquatic animals) vary at different depths and in different seasons.
- This is because factors that affect the photosynthesis of phytoplankton will vary:
  - There will be less light in winter and deeper in the water.
  - The temperature will be lower in winter and deeper in the water.
  - **Minerals** are used up towards the end of summer.

B–A*

- Food webs of marine organisms can provide useful information. The webs rely on different sources of food:
  - Most rely directly on green plants.
  - Others deeper in the ocean feed on dead material called **marine snow** that floats down.
  - Some rely on **bacteria**, deep in the ocean, acting as **producers**.

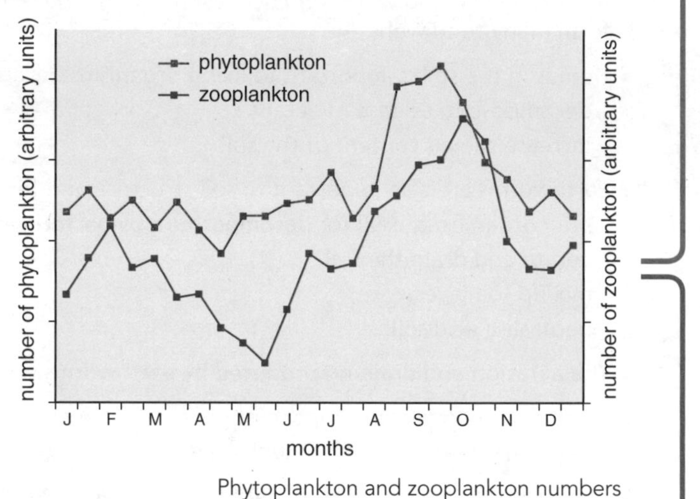

Phytoplankton and zooplankton numbers

Water pollution

D–C

- Sewage and fertiliser run-off can cause a process called **eutrophication** to occur. This involves the rapid growth of algae, which then all die and **decay**. This uses up oxygen, causing the death of animals because they are unable to **respire**.
- Some species of organisms are more sensitive to **pollution** than others and so they can be used as **biological indicators** for pH and oxygen.

B–A*

- Some chemicals such as PCBs and DDT can kill animals at the top of marine food chains. This is because the chemicals:
  - are toxic
  - do not break down quickly and so accumulate and become concentrated higher up the food chain
  - affect animals with a long lifespan, such as whales.

Remember!

DDT is a pesticide that caused poisoning due to ***accumulation*** higher up food chains. Fertilisers or sewage cause ***eutrophication***.

Improve your grade

Phytoplankton and zooplankton

Look at the graph above showing the numbers of zooplankton and phytoplankton.
Describe and suggest explanations for the changes in numbers between May and October.
AO2 [4 marks]

Enzymes in action

Enzymes in washing powder

D–C

- Biological washing powders often use **enzymes** such as:
  - **amylase**, to digest **carbohydrates** such as starch
  - **lipase**, to digest fat and remove fatty stains
  - **protease**, to digest protein and remove protein stains.
- Biological washing powders work best at moderate temperatures because this is the **optimum temperature** for enzymes to work.

B–A*

- After treatment with enzymes, the products of digestion are soluble and so will easily wash out of the clothes.
- Biological washing powders may not work in acidic or alkaline tap water because this is not the optimum for the enzymes and they might start to **denature**.

Remember!
Make sure that you know how enzymes work and understand the lock and key theory (covered in B3).

Enzymes and sweeteners

D–C

- Sucrose can be broken down by the use of an enzyme called **sucrase** (invertase).
- When sucrose is broken down by enzymes, the product is much sweeter, allowing the food industry to use less in food products.

B–A*

- Invertase converts sucrose into glucose and fructose.
- Glucose and fructose are sweeter than sucrose, so less has to be added to the food, lowering the cost and the energy content.

Lactase and immobilised enzymes

D–C

- Enzymes can be **immobilised** in gel beads by:
  - mixing the enzyme with **alginate**
  - dropping the mixture into calcium chloride solution.
- The immobilised enzymes produced are very useful in reactions. This is because:
  - the mixture does not become contaminated with the enzyme
  - they can be used in continuous flow processing.

B–A*

- Some people or animals are lactose intolerant because they cannot produce the enzyme lactase. This means that **bacteria** in the gut **ferment** lactose, which produces diarrhoea and wind.
- Milk can be treated for people who have lactose intolerance:
  - Immobilised lactase is used to convert lactose in milk into glucose and galactose.
  - Glucose and galactose can then be **absorbed** from the milk with no side effects.

EXAM TIP
Questions may refer to yogurt production. People with lactose intolerance can eat yoghurt. This is because bacteria have converted lactose in milk to lactic acid.

Improve your grade

Upset cats

Cats are lactose intolerant. Explain why it is not a good idea to feed them cow's milk and how the milk can be treated to make it suitable. *AO1* [4 marks]

Gene technology

Principles of genetic engineering

D–C

- **Genetic engineering** involves transferring a **gene** from one organism to another. The organism that receives the new gene is called a **transgenic organism**.
- The main stages in genetic engineering involve:
  - identifying and removing a desired gene from one organism
  - cutting open the **DNA** in another organism
  - inserting the new gene into the DNA
  - making sure that the gene works in the transgenic organism.
- The cutting and inserting of DNA is achieved using **enzymes** and often the transgenic organism can be **cloned** to produce identical copies.

B–A*

- The process of genetic engineering works because the genetic code is universal. This means that genes from one organism will produce the same protein in another organism.
- **Restriction enzymes** are used to cut open DNA. They leave several unpaired bases (single strands) on the cut end. This acts as a 'sticky end'.
- **Ligase** enzymes will join DNA strands because the 'sticky ends' on each cut section of DNA can join by complementary base pairing.

Remember!
The fact that all organisms have the same genetic code is used as good evidence for evolution by some scientists.

Genetically engineering bacteria

D–C

- **Bacteria** can be used in genetic engineering to produce human **insulin**. This involves:
  - cutting the gene for producing human insulin out of human DNA
  - cutting open a loop of bacterial DNA
  - inserting the insulin gene into the loop
  - inserting the loop into a bacterium.
- Many copies of the bacteria are cultured by **cloning** and large quantities of insulin are harvested.

B–A*

- The loops of DNA used in this process are called **plasmids**. They are found in the cytoplasm of bacteria and because they can be taken up by bacteria, they can be used as **vectors** for genes.
- To find out whether a bacterium has taken up a plasmid, an **assaying** technique is used:
  - Scientists add genes that make the bacteria resistant to **antibiotics**.
  - The bacteria are then flooded with the antibiotic by being grown on nutrient agar containing the antibiotic.
  - Scientists then choose the bacteria that survive.

DNA fingerprints

D–C

- **DNA 'fingerprints'** can be produced to identify individuals. They can be stored to help identify people who commit crimes and prove the innocence of others. However, some people are worried that they may be used for a variety of other reasons, such as assessing the likelihood of a person developing a disease. The information could be used as a reason to withhold life insurance.

B–A*

- The stages in the production of a DNA 'fingerprint' include:
  - extracting DNA from a sample, such as blood
  - cutting up or fragmenting the DNA using restriction enzymes
  - separating the fragments using electrophoresis
  - making the fragments visible using a radioactive probe.

Improve your grade

Enzymes and genetic engineering
Explain the role of enzymes in the process of genetic engineering. *AO1* [4 marks]

B6 Summary

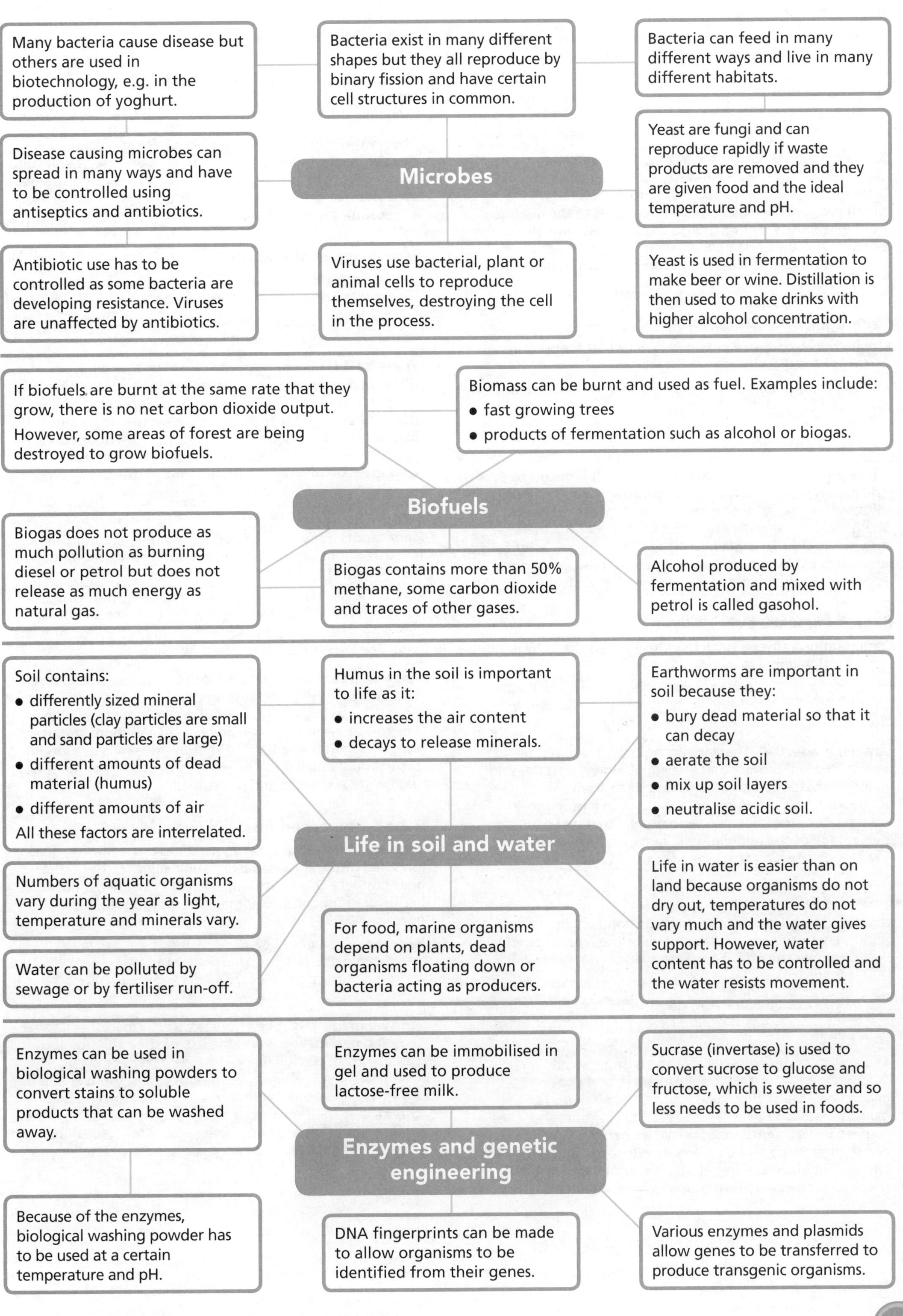

Many bacteria cause disease but others are used in biotechnology, e.g. in the production of yoghurt.
Bacteria exist in many different shapes but they all reproduce by binary fission and have certain cell structures in common.
Bacteria can feed in many different ways and live in many different habitats.
Disease causing microbes can spread in many ways and have to be controlled using antiseptics and antibiotics.
Microbes
Yeast are fungi and can reproduce rapidly if waste products are removed and they are given food and the ideal temperature and pH.
Antibiotic use has to be controlled as some bacteria are developing resistance. Viruses are unaffected by antibiotics.
Viruses use bacterial, plant or animal cells to reproduce themselves, destroying the cell in the process.
Yeast is used in fermentation to make beer or wine. Distillation is then used to make drinks with higher alcohol concentration.
If biofuels are burnt at the same rate that they grow, there is no net carbon dioxide output.
However, some areas of forest are being destroyed to grow biofuels.
Biomass can be burnt and used as fuel. Examples include:
• fast growing trees
• products of fermentation such as alcohol or biogas.
Biofuels
Biogas does not produce as much pollution as burning diesel or petrol but does not release as much energy as natural gas.
Biogas contains more than 50% methane, some carbon dioxide and traces of other gases.
Alcohol produced by fermentation and mixed with petrol is called gasohol.
Soil contains:
• differently sized mineral particles (clay particles are small and sand particles are large)
• different amounts of dead material (humus)
• different amounts of air
All these factors are interrelated.
Humus in the soil is important to life as it:
• increases the air content
• decays to release minerals.
Earthworms are important in soil because they:
• bury dead material so that it can decay
• aerate the soil
• mix up soil layers
• neutralise acidic soil.
Life in soil and water
Numbers of aquatic organisms vary during the year as light, temperature and minerals vary.
Life in water is easier than on land because organisms do not dry out, temperatures do not vary much and the water gives support. However, water content has to be controlled and the water resists movement.
For food, marine organisms depend on plants, dead organisms floating down or bacteria acting as producers.
Water can be polluted by sewage or by fertiliser run-off.
Enzymes can be used in biological washing powders to convert stains to soluble products that can be washed away.
Enzymes can be immobilised in gel and used to produce lactose-free milk.
Sucrase (invertase) is used to convert sucrose to glucose and fructose, which is sweeter and so less needs to be used in foods.
Enzymes and genetic engineering
Because of the enzymes, biological washing powder has to be used at a certain temperature and pH.
DNA fingerprints can be made to allow organisms to be identified from their genes.
Various enzymes and plasmids allow genes to be transferred to produce transgenic organisms.

Page 4 Heart disease

Excess saturated fat or excess salt in the diet can increase the risk of heart disease. Explain why. *AO1* [4 marks]

If there is too much saturated fat in the diet, cholesterol builds up in the wall of the arteries, so the heart gets less blood. This means that the cells die. Too much salt will increase the blood pressure, making heart disease more likely.

Answer grade: D–C. There is a correct link between saturated fats in the diet and cholesterol build up in arteries. There is also correct reference to salt raising the blood pressure. However, the answer refers to the heart and not heart muscle. For full marks, explain that high blood pressure is likely to damage the walls of the blood vessels. This produces blood clots or a thrombosis, which blocks the blood vessels.

Page 5 EAR for protein

Explain the importance of knowing your EAR for protein. *AO1/2* [4 marks]

The EAR is the estimated average requirement in your diet each day.

The protein is needed for growth.

Too much or too little will cause a problem.

Answer grade: D–C. The answer states what is meant by EAR and knows protein is important for growth. For full marks, explain that too little protein can cause kwashiorkor and that the body cannot store excess protein. For full marks, answer would also explain that the EAR is only an estimated amount for the average person and that a fast-growing teenager would need a greater amount of protein.

Page 6 Immunisation

Immunisation carries a small risk. Despite this risk, why is immunisation important? *AO1* [4 marks]

Immunisation will protect the individual against a particular disease. It will also prevent many other people getting that disease.

Answer grade: D–C. The answer correctly states that being immunised will protect that individual. However, it does not mention that the small risk of side effects from immunisation is offset by the avoidance of a potentially fatal disease. For full marks, also explain that when a high percentage of the population is immunised there is a very low risk of infection from that disease.

Page 7 Vision

In old age, muscles lose their ability to contract and relax quickly. Ligaments become less flexible. The lens becomes less elastic. Explain the effects of these changes on vision. *AO2* [4 marks]

The eye muscles and ligaments cannot work properly. The eye now cannot accommodate. This will affect vision, especially in older people.

The answer states correctly that the changes would affect accommodation, but does not explain it, and does not explain how vision would be affected. For full marks, explain that if the ciliary muscles cannot contract and if the suspensory ligaments are not flexible then the lens would not change shape and accommodate (change focus) for near and distant objects. A less elastic lens would also result in the inability of the lens to change shape and focus.

Page 8 Affects of alcohol

Drinking alcohol increases the risk of accidents. Explain why. *AO2* [4 marks]

Alcohol is a depressant drug. It affects reaction time to danger. A slow reaction time will increase the risk of an accident. Alcohol affects nerve transmission across synapses.

The answer correctly links up alcohol being a depressant drug with slow reaction times and accident risk. It also states that alcohol has an effect on the nerve transmission across synapses but did not say what effect. Details of the blocking of receptor molecules should be included for full marks.

Page 9 Types of diabetes

Joe has Type 1 diabetes and Charlie has Type 2 diabetes. Explain why they require different treatments. *AO1* [4 marks]

Joe needs regular insulin doses, whereas Charlie can regulate his diet. Both treatments regulate blood sugar levels.

There is no explanation of the causes of the two types of diabetes. For full marks, explain that Joe's pancreas is unable to make any insulin and so needs regular insulin doses. Charlie's body is producing some insulin so if he lowers his sugar/carbohydrate intake he will not need so much insulin. He will also need to match his sugar/carbohydrate intake with his amount of exercise.

Page 10 Phototropism

Plant shoots grow straight upwards in the dark but will grow towards a light source. Explain how and why they do this. *AO1* [4 marks]

Plant shoots will grow upwards and towards the light so the plant's leaves will photosynthesise faster. Plant shoots are called positively phototropic because they grow towards the light. The curvature is controlled by auxins.

Answer grade: D–C. There is a correct link between light and photosynthesis and stating that the curvature towards light is caused by auxins. For full marks, state that a higher amount of auxin in the shaded part of the shoot results in longer cells in that area, and that in the dark there would be an equal distribution of auxin in the shoot, so the shoot would grow straight and not curved.

Page 11 Inherited disorders

Rabeena finds out that her mother has cystic fibrosis, an inherited disorder caused by a faulty allele (c). Her father does not have this faulty allele. Explain why Rabeena hopes her future husband will not be heterozygous for cystic fibrosis. *AO2/3* [4 marks]

Rabeena will not have cystic fibrosis but she will be heterozygous (Cc). If her future husband is also heterozygous for cystic fibrosis, some of their children could have cystic fibrosis.

Answer grade: D–C. The answer correctly identifies Rabeena as being heterozygous and the risk of their children having cystic fibrosis. For full marks, include details of possible children. If her husband was heterozygous, there would be a 1 in 4 chance of a child having cystic fibrosis, an equal chance of a child carrying the faulty allele and a 1 in 4 chance of not carrying any faulty alleles for cystic fibrosis. This could be shown in a Punnett square.

Page 13 Classifying newly discovered organisms

Two similar types of animals have been discovered living close together in a jungle. Describe how scientists could find out how closely related the two animals are. *AO2* [3 marks]

The scientists could breed the animals together to see if they are the same species.

Answer grade: D–C. The answer gives a method but does not explain the possible outcomes of the cross. For full marks, the student would need to state that they would be the same species if offspring were fertile. The base sequence in their DNA could also be compared.

Page 14 Pyramids of biomass

Explain one advantage and one disadvantage of using pyramids of biomass to show feeding relationships. *AO2* [3 marks]

Pyramids of biomass are useful because they are always pyramids, unlike pyramids of numbers. However, it is hard to get the information to draw them.

Answer grade: D–C. The answer includes an advantage and a disadvantage but neither is explained properly. The explanation should say that pyramids of biomass are always pyramids because the energy decreases at each trophic level. They therefore give an idea of the energy loss. However, they involve measuring dry mass, which is a destructive process.

Page 15 The nitrogen cycle

Explain how nitrogen in a protein molecule in a dead leaf can become available again to a plant. *AO1* [4 marks]

The protein is broken down by decomposers such as bacteria and fungi in the soil. Nitrates are produced, which are taken up by the roots of plants.

Answer grade: D–C. The answer states that decomposers break down proteins and that plants take up nitrates. The intermediate steps are missing, however. For full marks, state that: bacteria and fungi in the soil decompose the dead leaves, they convert protein into ammonia, which is then converted into nitrates by nitrifying bacteria and the nitrates can be taken up by the roots of plants.

Page 16 Niches

The scientist Gauss put forward a theory that said organisms of two different species cannot share an identical ecological niche.

(a) Explain what is meant by the term 'ecological niche'. *AO1* [2 marks]

(b) Suggest why Gauss said that two species cannot share the same niche. *AO2* [2 marks]

(a) The term ecological 'niche' describes what an organism eats.

(b) The two organisms would be eating the same food.

Answer grade: D–C. What an organism eats is only part of the definition of ecological niche. For full marks, state that the term describes where an organism lives and its role in the ecosystem and that if two organisms had the same niche, then one would out-compete the other.

Page 17 Living in hot, dry conditions

An elephant has a large body, large ears, skin with few hairs and the ability to produce concentrated urine. Explain which of these features are advantages or disadvantages when living in hot, dry areas. *AO2* [4 marks]

The large ears will help the elephant lose heat. The lack of hair will stop the animal over-heating as it will not prevent heat loss. Producing a concentrated urine will help conserve water.

Answer grade: D–C. The answer gives some advantages with some explanation but has not given any disadvantages. For full marks, discuss the large size of the elephant, which gives it a small surface area to volume ratio, making it harder to lose heat. However, the large ears help to increase the surface area. Thin hair will allow heat loss but does not give much insulation from the sun's rays.

Page 18 Explanations for evolution

Human ancestors had more hair than modern humans. This could be explained by saying that scratching the skin due to parasites has gradually over many generations made some of the hair fall out.

(a) Explain why this explanation uses Lamarck's ideas. *AO2* [2 marks]

(b) How might Darwin's ideas be used to explain why modern humans have less hair? *AO2* [2 marks]

(a) This suggestion uses the idea that characteristics can change.

(b) Humans with less hair are less likely to get parasites and so this is an advantage.

Answer grade: D–C. The answer to (a) does not say that the characteristics are acquired during the organism's life and then passed on. In part (b) the advantage is a reasonable idea, but the ideas of variation and being able to survive to pass on the genes for shorter hair must also be included.

Page 19 Population and pollution

Explain the reasons for the increase in carbon dioxide levels in the atmosphere and explain why people are concerned about this. *AO1* [4 marks]

More carbon dioxide is released from the increased burning of fossil fuels such as coal and oil. People are worried that this might cause global warming.

Answer grade: D–C. The answer gives a reason for the increase of carbon dioxide and a possible consequence but neither is fully explained. For full marks the student should also say why fossil fuels are being burned and explain some of the possible consequences of global warming.

Page 20 Saving endangered species

The Hawaiian goose is only found on the islands of Hawaii. In the mid-1900s only about 30 were left alive. The space on the islands is restricted and a number of animals to the islands have been introduced. Write about the problems facing scientists who are trying to save the goose from extinction in Hawaii. *AO2* [4 marks]

There are not many geese left to breed with and there is also little room for them to live on the islands. The animals on the island may harm them.

Answer grade: D–C. The answer highlights the small number of geese that are left and two of the problems on the island. For full marks the student should also refer to the lack of genetic variation left in the geese, and to the lack of suitable habitats, possible predation and competition from other animals.

Page 22 Protein synthesis

Describe where and how proteins are coded for and made. *AO1* [6 marks]

Proteins are coded for by DNA. They are made in the cytoplasm on ribosomes.

Answer grade: D–C. Sentence 1 correctly states that DNA codes for proteins but does not say how. Sentence 2 correctly gives the site of production but not how the information reaches there. For full marks, include details of how the order of bases codes for the order of amino acids and discuss the function of mRNA.

Page 23 Mutations and enzymes

A mutation can occur in a gene that codes for an enzyme. Explain how a mutation could lead to an enzyme failing to work properly. *AO1* [6 marks]

The mutation could change the bases in the DNA of the gene. This could mean that the protein is made with a different shape.

Answer grade: D–C. Sentence 1 correctly states that the bases might change but does not explain how this changes the protein. Sentence 2 suggests that the shape changes but does not explain how this changes the function. For full marks, a change in the order of amino acids needs to be stated and some reference to a change in shape preventing the substrate fitting into the active site.

Page 24 Respiration and enzymes

An experiment was set up to measure the oxygen uptake of insects. The data shows that the maximum uptake was at about 35 °C. Above and below that temperature less oxygen was used.

Explain what the oxygen is used for and why it is used fastest at 35 °C. *AO1* [2 marks], *AO2* [2 marks]

The oxygen is used for respiration. 35 °C is closest to body temperature and so this is the optimum temperature for respiration.

Answer grade: D–C. Sentence 1 links oxygen to respiration but does not say which type. Sentence 2 uses the term 'optimum' but does not explain why there is an optimum. For full marks, the word 'aerobic' should be used to describe respiration and the role of enzymes should be discussed.

Page 25 Meiosis and mitosis

The two processes mitosis and meiosis occur in the human body. Compare each process, writing about where they occur and any differences in the process. *AO1* [6 marks]

Mitosis happens all over the body but meiosis makes gametes. They both make new cells but in mitosis they have the same number of chromosomes and in meiosis they have half.

Answer grade: D–C. Sentence 1 correctly states where mitosis occurs but not where meiosis happens. Sentence 2 compares the outcome but not the process. For full marks the answer should say that meiosis occurs in the sex organs. Also, in mitosis there is one cell division, during which the copies of the chromosomes separate. In meiosis, there are two divisions. In the first, the chromosomes pair up and separate. In the second, copies move to the opposite poles of the cell.

Page 26 Valves

Valves are found in veins, at the start of the arteries leaving the heart and in the heart. Write about the importance of these different valves. *AO1/2* [2 marks]

The job of valves is to stop blood flowing backwards. In the veins it keeps it moving back to the heart. In the heart it makes sure that it does not flow back into the atria.

Answer grade: D–C. The answer explains the job of valves in general but does not say why they are needed in veins. Also, there is not enough detail about the valves in the heart and arteries. For full marks, include an explanation involving the low pressure of the blood in veins and the role of the semilunar valves in stopping the blood flowing back into the heart when the heart muscle relaxes.

Page 27 A plant growth curve

Katie wants to plot a growth curve for a broad bean plant using dry mass. Given 100 broad bean seeds, explain how you would collect the data to plot the graph and explain why she wants to plot dry mass. *AO1* [5 marks]

I would plant the seeds and every week dry out one of the seedlings and weigh it. Dry mass gives the best measure of growth.

Answer grade: D–C. Sentence 1 states that the 100 seeds are sampled every month but does not give details of how they are dried. Sentence 2 does not give a reason for why dry mass is preferable. For full marks, explain that the seedlings have to be dried in an oven at about 80 °C, which prevents burning. Also, dry mass measures permanent growth in all directions, not just changes in water content.

Page 28 Spider silk

Spider silk is very strong and could be very useful in industry. Goats have now been produced that make spider silk in their milk.

Describe how this could be done and suggest reasons why this method of production might be more useful.
AO2 [4 marks]

The spider silk DNA is put into goats. This is useful because the silk is easier to collect from the milk.

Answer grade: D–C. The answer briefly describes the process and gives one possible advantage. For full marks, it would need to say that the spider gene for silk is isolated and transferred. Also, more than one reason is needed, such as a comparison of the quantity made.

Page 29 Cloning plants

A garden centre wants to sell an attractively coloured geranium plant. They decide to produce many clones of the plant using tissue culture.

(a) What are the possible disadvantages of this method of reproduction? *AO1* [2 marks]

(b) Why is this method not possible for producing goldfish for garden ponds? *AO2* [2 marks]

(a) The plants might all die at once.

(b) Because animals cannot be cloned in this way.

Answer grade: D–C. (a) is incomplete because it does not explain why. For full marks, it should state that the lack of genetic variation (and so the susceptibility to disease) is important. (b) is also incomplete because no reason is given. It should state the lack of ability of animal cells to differentiate.

Page 31 Distribution of animals

Rick wants to find out about the zonation of different types of limpets down a sea shore. Explain:

(a) how he should do this

(b) what can cause this zonation. *AO1/2* [5 marks]

(a) He puts a long line down the sea shore and counts the limpets at different stages.

(b) The limpets at the top of the shore will be more exposed at the top of the shore and may not survive.

Answer grade: D–C. For full marks, (a) should describe using a quadrat at measured intervals and state that different species should be investigated. (b) should also say that the abiotic factors would have a different effect on different limpet species.

Page 32 Exchange of gases in plants

When gas exchange in plants is analysed, they seem to respire only at night. Explain why. *AO2* [4 marks]

A plant will carry out respiration during both day and night. Plants respire by taking in oxygen and releasing carbon dioxide. The oxygen is used to release energy from glucose.

Answer grade: D–C. For full marks, include details of photosynthesis taking place in the light, using up carbon dioxide and producing oxygen. Also, explain that the rate of gas exchange in photosynthesis is more than that in respiration, so respiration in plants during daylight is difficult to isolate and measure.

Page 33 Absorption of light

Look at the graph.

(a) Which parts of the light spectrum are i) reflected and ii) used by leaves? *AO2* [2 marks]

(b) Explain why a leaf has many different pigments. *AO1* [2 marks]

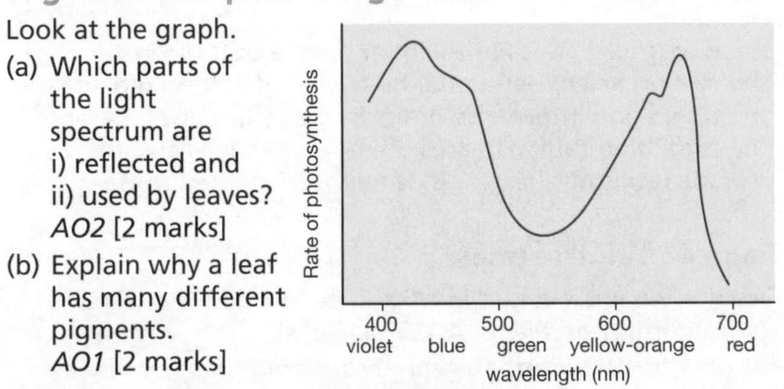

(a) i) A wide range of colours are used by leaves. The middle part of the spectrum is reflected.

ii) The middle part is not used by leaves.

(b) The leaf has many different pigments so it can be a variety of colours. It can also use a lot of different colours in the spectrum.

Answer grade: D–C. (a) i) gives a vague answer ('middle part') instead of quoting a colour (green) or actual numbers. In ii) needs to state what *is* used and the colours (violet/blue and yellow/orange) or numbers in nm. For full marks in (b), quote names of pigments as well as using the word 'photosynthesis'.

Page 34 Osmosis in plant and animal cells

Describe and explain the effects of water entry into plant and animal cells. *AO1/2* [5 marks]

When water enters plant and animal cells the turgor pressure increases, helping to maintain the cell shape. Because animal cells do not have a cell wall, they can swell up if too much water enters.

Answer grade: D–C. For full marks, Sentence 1 should also state that the process involves osmosis, and describe it and how it works. The answer states that an animal cell would swell up, but should also explain that in the absence of a cell wall, the cell could burst and die (lysis).

Page 35 Transport in plants

Marram grass grows in exposed sand dunes. It has narrow leaves, stomata sunk in pits, many hairs on its leaves and leaves that can curl up into a tube. Explain how these adaptations help it to survive. *AO2* [4 marks]

As Marram grass grows in exposed places it must reduce its water loss, or its roots will not be able to take up enough water to maintain turgor and carry out photosynthesis. All these adaptations will reduce its water loss and help it to survive.

Answer grade: D–C. The answer is good, as it explains why Marram grass will have problems in retaining enough water. It also makes a link between reducing water loss and survival. For full marks, explain how each of the adaptations would decrease water loss, e.g. narrow leaves have a smaller surface area and fewer stomata than broad leaves.

Page 36 Mineral uptake

Look at the graph.

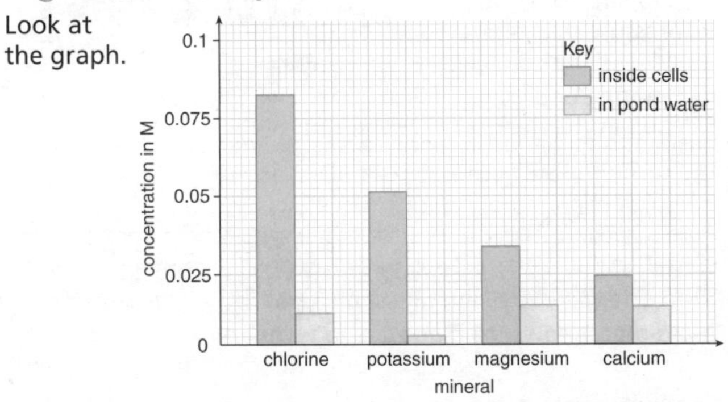

What conclusions can be made from this data? *AO3* [4 marks]

The graph shows that chlorine, potassium, magnesium and calcium are taken up by the algae. Chlorine is taken up the most and calcium the least.

Answer grade: D–C. For full marks, the answer should also say that the data shows that minerals are taken up against a concentration gradient and in different amounts. This indicates a system of active transport, a system of carriers using energy to transport minerals into the cell's cytoplasm.

Page 37 Growing mushrooms

Lynn wants to grow mushrooms (a fungus). Explain what conditions she should provide for optimum growth. *AO1/2* [3 marks]

Lynn should grow them in soil with plenty of water, minerals, a reasonable temperature and enough oxygen.

Answer grade: D–C. The question asked for an explanation. For full marks, state that a temperature of 25 °C is required for optimum respiratory rate; water is required by a saprophyte such as mushrooms for extracellular digestion; plenty of oxygen is required for aerobic respiration for optimum growth and reproduction.

Page 38 Hydroponics

Look at the diagram of a hydroponics system. Explain how this system is useful for growing lettuce in glasshouses. *AO1/2* [4 marks]

Lettuce are small plants and do not need support in a hydroponics system. The system supplies the water, minerals and air the lettuce needs to grow. The system does not use soil.

Answer grade: D–C. For full marks, include details about being able to: grow a large number of lettuces in a small space (can regulate and recycle the amount of minerals supplied for optimum growth); use the system in glasshouses and avoid the crop being eaten by animals; control climate conditions, so grow a succession of crops all year round; and to grow lettuce in areas of poor soil.

Page 40 Bones

Ruth wants to know if she has stopped growing. Explain how an x-ray could provide an answer. *AO1/2* [4 marks]

Long bones start off being cartilage. When calcium salts are deposited in the bones, the bones become much harder. An x-ray of her bones will show if some cartilage remains.

Answer grade: D–C. For full marks, state that phosphorus is also deposited; where the remaining cartilage would be; and name the process (ossification). Link the presence of cartilage with the potential for further growth.

Page 41 Pressure changes in blood vessels

Look at the graph showing pressure changes in blood vessels.

(a) Explain what happens to blood pressure in arteries. *AO2* [2 marks]

(b) Describe and explain the general trend shown in the graph. *AO2* [2 marks]

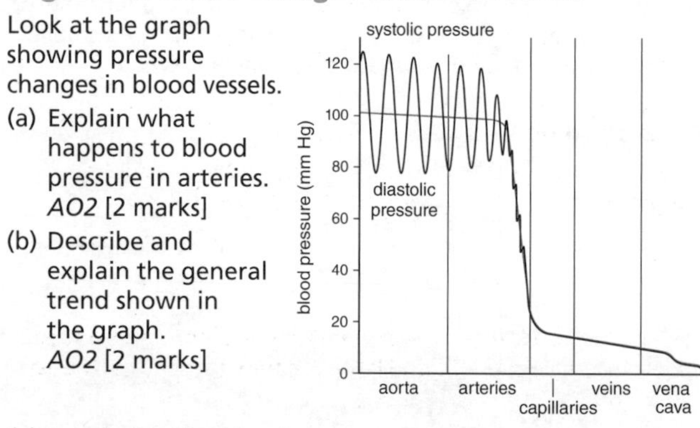

(a) In arteries, the blood pressure varies in a regular pattern.

(b) As blood flows from arteries to veins and then to capillaries, the pressure drops until it is very low in capillaries.

Answer grade: D–C. In (a), state that as the heart goes through the cardiac cycle, the contraction of the powerful muscle of the left ventricle puts the blood under high pressure. This pressure drops as the ventricle refills with blood, ready to be distributed around the body. In (b), state that as the blood gets further from the heart the pressure decreases until the blood is flowing in capillaries with walls only one cell thick.

Page 42 Blood groups

Look at the diagram of blood group agglutinins. Explain why people with blood type AB can donate blood only to people with the same blood type. *AO2/3* [4 marks]

There are four main blood groups called A, B, AB and O. The groups depend on the absence or presence of agglutinins. Blood type AB has antigens A and B but no A or B antibodies, whereas blood groups A and B have antibodies.

Answer grade: D–C. For full marks, explain why AB blood should not be given to people with another blood type. All the other types contain antibodies, which would react with AB blood, causing it to agglutinate.

Page 43 Spirometer recording

Look at the diagram showing a spirometer recording of a healthy adult.

(a) Work out this person's
(i) total lung capacity and
(ii) vital capacity.
AO2 [2 marks]

(b) Spirometer readings are not usually taken during an asthma attack. If they were, suggest what changes in recordings would be expected and explain why they occur. *AO3* [2 marks]

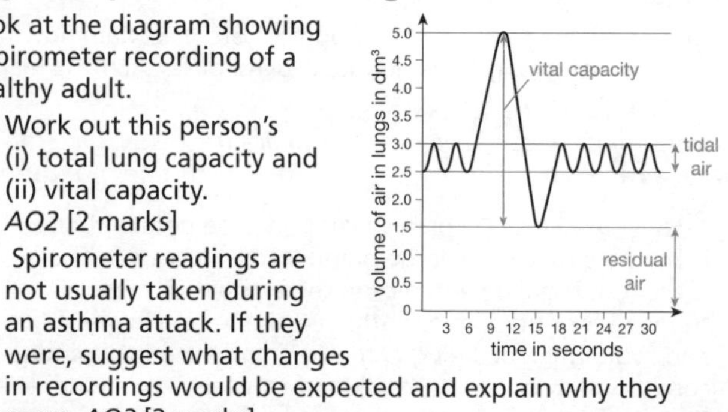

(a) The total lung capacity is 5 and the vital capacity is 3.5.

(b) I expect the total lung capacity and vital capacity will be reduced if the person has asthma because the airways would be constricted.

Answer grade: D–C. In (a), the total lung capacity is not reduced as it is the airways that are constricted, not the air sacs and alveoli. In (b), the amount of tidal air would decrease and the amount of residual air would increase because of these constrictions. The rate of air flow in and out of the lungs would also decrease, so the time would be extended.

Page 44 Gall bladder

Jason's gall bladder has been removed in an operation. Explain why he may have to change his diet. *AO1/2* [3 marks]

The gall bladder stores bile, which is involved in the digestion of fats. It is a small bag attached to the liver. Jason should not eat too many fats because he can't digest them.

Answer grade: D–C. It is incorrect to say that Jason will be unable to digest fats. Bile emulsifies fats and therefore helps fat digestion by creating a larger surface area. This does not mean that lipase enzymes cannot work, just that fat digestion may not be efficient.

Page 45 Carbon dioxide concentration

Sonya uses her exercise machine. Her increased rate of breathing will get more oxygen into her body. Explain another reason why her breathing rate will increase. *AO1/2* [4 marks]

Sonya will use up oxygen and produce carbon dioxide in respiration so the levels of carbon dioxide in her body will increase. Her body will detect the increase and increase her rate of breathing.

Answer grade: D–C. For full marks, state that receptors in the carotid artery detect the high level of carbon dioxide in the blood and inform the brain via nerve impulses. Raising the breathing rate increases the rate at which carbon dioxide (toxic at high levels) is removed by the lungs.

Page 46 IVF treatment

Describe what is involved in IVF and explain how it can solve some infertility problems. *AO1/2* [4 marks]

IVF (in vitro fertilisation) is the fertilisation of an egg by a sperm outside the human body. An egg is taken from a female and mixed with sperm from a male. This helps an infertile couple to have a child.

Answer grade: D–C. Mixing the egg and sperm is useful when ovulation is erratic, oviducts are blocked, or only a small amount of sperm is produced. It will not increase the chances of fertilisation where there is no egg or sperm production or an inability to carry an unborn baby to term.

Page 47 Transplant survival rates

Look at the graph showing survival rates from transplants in the UK in 2010. Suggest why different organ transplants have different survival rates. *AO2* [3 marks]

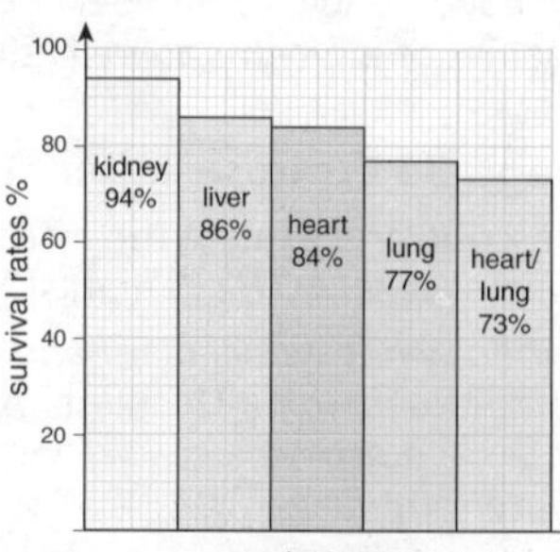

Transplanting some organs requires more extensive surgery than others.

Answer grade: D–C. For full marks, state which transplants showed the highest/lowest survival rates and link them to how extensive the surgery would be. The age of the recipients is also a factor.

Page 49 Viruses

Although they do not feed, viruses are usually described as parasites.

Explain why this is. *AO2* [4 marks]

This is because they damage other cells by injecting their genetic material into them and making them split open.

Answer grade: D–C. The answer does not say how the virus gains from doing this. For full marks, state that the virus uses the 'host' cell to make more copies of itself before the cell is destroyed.

Page 50 Barbeque time

In the summer, when more food is cooked on barbeques, more people get food poisoning (e.g. from *Salmonella*). Explain why this is and what can be done to prevent the problem. *AO2* [3 marks]

This is because the food at barbeques might not have been prepared as carefully as food is indoors. Everybody should wash their hands.

Answer grade: D–C. Washing hands helps, but the key point is that food needs to be cooked thoroughly to kill all microbes.

Page 51 Yoghurt making

Brenda decides to make her own yoghurt by adding a small amount of yoghurt that she bought to some milk. Her friend Sarah says that this will not work because the bought yoghurt had been pasteurised before it was packaged.

Explain why Sarah was correct. *AO2* [3 marks]

Sarah was correct because pasteurising involves heating up the yoghurt to 78°C, which sterilises it, killing all the bacteria.

Answer grade: D–C. The conditions for pasteurisation are correct but this will not sterilise the yoghurt. However, pasteurisation does kill the Lactobacillus bacteria needed to turn milk into yoghurt. This is why Brenda's process will not work.

Page 52 Biogas use

Discuss the advantages and disadvantages of biogas compared to diesel and natural gas. *AO1* [4 marks]

Biogas has a major advantage over diesel and natural gas because it is not a fossil fuel. It therefore does not release greenhouse gases.

Answer grade: D–C. Biogas does release greenhouse gases but the key point is that it releases them at the same rate that they are taken in by plants. For full marks, state that biogas does not produce particulates but diesel does. Also, give a disadvantage of biogas, such as the fact that it releases less energy than gas when it is burnt.

Page 53 Soil analysis

A student analyses two different soils. Here are his results.

Soil	pH	Air content	Humus content
A	5.8	13.2	5.4
B	7.2	18.5	12.8

Which soil is likely to contain more earthworms? Justify your answer. *AO3* [2 marks], *AO2* [2 marks]

Soil B. This is because it has the highest pH and the highest air content and highest humus content.

Answer grade: D–C. Soil B and the reasons are correct but not explained. For full marks, explain that earthworms neutralise acidic soils; make burrows (which aerates the soil); and drag leaves down into the soil.

Page 54 Phytoplankton and zooplankton

Look at the graph showing the numbers of zooplankton and phytoplankton. Describe and suggest explanations for the changes in numbers between May and October.
AO2 [4 marks]

The numbers of both organisms increase from May but then start to decrease in October. This is because there is more light available in the summer and so more photosynthesis. The light starts to drop in October.

Answer grade: D–C. The answer correctly describes the growth pattern, but should also state that the change in zooplankton numbers occurs after the change in numbers of phytoplankton. Increasing or decreasing phytoplankton numbers affects zooplankton numbers due to food availability.

Page 55 Upset cats

Cats are lactose intolerant. Explain why it is not a good idea to feed them cow's milk and how the milk can be treated to make it suitable. *AO1* [4 marks]

The cats will not be able to digest the milk and it will make them ill. To make it suitable it should be treated with immobilised enzymes.

Answer grade: D–C. It is true that cats cannot digest milk but it is the lactose that they cannot digest. For full marks, also state that the immobilised enzyme used in the treatment is lactase.

Page 56 Enzymes and genetic engineering

Explain the role of enzymes in the process of genetic engineering. *AO1* [4 marks]

Enzymes are used to make chemical reactions happen faster. They can cut up sections of DNA and also join sections together.

Answer grade: D–C. The role of enzymes is correct but not specific enough. For full marks, state that restriction enzymes cut DNA and leave sticky ends. They are used to cut out the gene and to make a cut in the plasmid. Ligase enzymes can then be used to join the sticky ends and so enable insertion of the gene into the plasmid.

Understanding the scientific process

As part of your Biology assessment, you will need to show that you have an understanding of the scientific process – How Science Works.

This involves examining how scientific data is collected and analysed. You will need to evaluate the data by providing evidence to test ideas and develop theories. Some explanations are developed using scientific theories, models and ideas. You should be aware that there are some questions that science cannot answer and some that science cannot address.

Collecting and evaluating data

You should be able to devise a plan that will answer a scientific question or solve a scientific problem. In doing so, you will need to collect data from both primary and secondary sources. Primary data will come from your own findings – often from an experimental procedure or investigation. While working with primary data, you will need to show that you can work safely and accurately, not only on your own but also with others.

Secondary data is found by research, often using ICT – but do not forget books, journals, magazines and newspapers are also sources. The data you collect will need to be evaluated for its validity and reliability as evidence.

Presenting information

You should be able to present your information in an appropriate, scientific manner. This may involve the use of mathematical language as well as using the correct scientific terminology and conventions. You should be able to develop an argument and come to a conclusion based on the recall and analysis of scientific information. It is important to use both quantitative and qualitative arguments.

Changing ideas and explanations

Many of today's scientific and technological developments have both benefits and risks. The decisions that scientists make will almost certainly raise ethical, environmental, social or economic questions. Scientific ideas and explanations change as time passes and the standards and values of society change. It is the job of scientists to validate these changing ideas.

How science ideas change

From the information you have learnt, you will know that science is a process of developing, then testing, theories and models. Scientists have been carrying out this work for many centuries and it is the results of their ideas and trials that have provided us with the knowledge we have today.

However, in the process of developing this knowledge, many ideas were put forward that seem quite absurd to us today.

During the Middle Ages, the 'miasma theory' explained how diseases were caused. Miasma was thought to be a poisonous vapour present in the air. This vapour was said to contain particles of decaying matter that created a foul smell. The name of the killer disease malaria is derived from the Italian *mala*, meaning 'bad' and *aria*, meaning 'air'.

In the nineteenth century, England was undergoing a rapid expansion of industrialisation and urbanisation. This created many foul-smelling and filthy neighbourhoods, which were focal points for disease. By improving housing, cleanliness and sanitation, levels of disease fell. This fall in the level of disease supported the miasma theory.

In 1854, John Snow confirmed a cholera outbreak in London as originating from a water pump. Within ten years, Louis Pasteur was suggesting that it was the presence of germs in substances such as milk and meat that caused them to go off quickly. He was able to remove the germs by a process we now call *pasteurisation*. This process is still used to protect perishable foodstuffs today.

Reliability of information

It is important to be able to spot when data or information is presented accurately. Just because you see something online or in a newspaper, does not mean that it is accurate or true.

Think about what is wrong in this example, based on a newspaper report. Look at the answer at the bottom of the page to check that your observations are correct.

MP challenges Health Minister

The Health Minister was challenged in the House by local MP, Ralph Stag, following this week's publication of figures for the number of adult males affected by oscillatory plumbosis. Mr Stag cited this year's "sudden and dramatic increase" in the number of sufferers and demanded swift action to discover the causes of this "worrying trend".

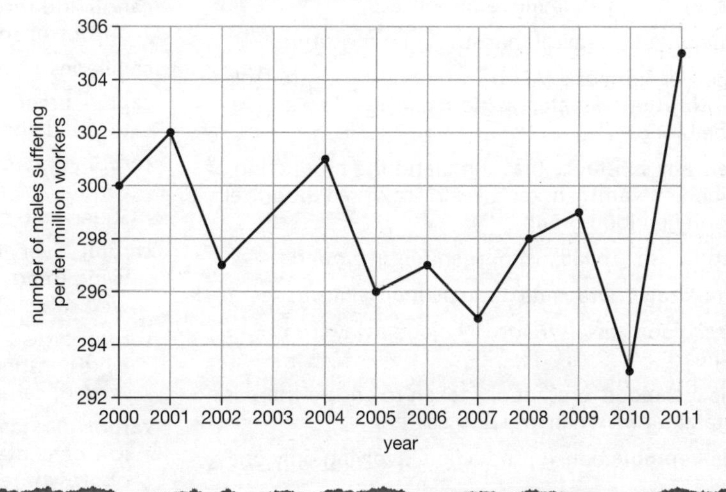

Answer

The *y* axis does not start at zero and so a change from 293 to 305 individuals in one year is a difference of only 12 people per ten million workers. This is unlikely to be significant and is not really a 'sudden and dramatic increase'.

Glossary

A

absorbed taken in

accommodation the eye's ability to change focus

acid rain rain water which is made more acidic by pollutant gases

acrosome part of the sperm that contains enzymes

active immunity you have immunity if your immune system recognises a pathogen and fights it

active site the place on an enzyme where the substrate molecule binds

active transport in active transport, cells use energy to transport substances through cell membranes against a concentration gradient

adaptations features that organisms have to help them survive in their environment

adrenaline hormone produced by adrenal gland

aerating increasing the amount of air in soil

aerobic respiration respiration that involves oxygen

agglutination when red blood cells group together

alcohol substance made by the fermentation of yeast

alginate chemical used to immobilise enzymes

allele inherited characteristics are carried as pairs of alleles on pairs of chromosomes. Different forms of a gene are different alleles

amino acids small molecules from which proteins are built

amniocentesis test during pregnancy for foetal abnormalities

amylase enzyme that digests starch

anaerobic respiration respiration without using oxygen

antagonistic muscles pair of muscles working together, as one contracts the other relaxes

antibiotic therapeutic drug acting to kill bacteria which is taken into the body

antibody protein normally present in the body or produced in response to an antigen which it neutralises, thus producing an immune response

anti-coagulant chemical that stops blood clotting

anti-diuretic hormone (ADH) hormone which controls re-absorption of water in kidneys (controls water levels in the blood)

antigen any substance that stimulates the production of antibodies – antigens on the surface of red blood cells determine blood group

antiseptic used to kill microorganisms in wounds

antiviral drug therapeutic drug acting to kill viruses

arteries blood vessels that carry blood away from the heart

aseptic technique precautions taken to ensure there is no contamination when growing bacteria

asexual reproduction reproduction involving only one parent

assaying technique used in genetic engineering to find out if bacteria have taken up the genes

asthma condition where airways are inflamed and constricted

ATP molecule used to store energy in the body

auxin a type of plant hormone

axon part of neurone that carries nerve impulse

B

bacteria single-celled micro-organisms which can either be free-living organisms or parasites (they sometimes invade the body and cause disease)

balanced symbol equation a symbolic representation showing the kind and amount of the starting materials and products of a reaction

binary fission reproduction in bacteria

binocular vision the ability to maintain visual focus on an object with both eyes, creating a single visual image

binomial system the scientific way of naming an organism

biodiversity range of different living organisms in a habitat

biofuels fuels made from plants – these can be burned in power stations

biogas biofuel containing methane

biological catalyst molecules in the body that speed up chemical reactions

biological control a natural predator is released to reduce the number of pests infesting a crop

biological indicators organisms living in water: their presence or absence tells scientists how polluted water is

biomass waste wood and other natural materials which are burned in power stations

blind trial a drugs trial where volunteers do not know which treatment they are receiving

blood pressure force with which blood presses against the walls of the blood vessels

blood sugar level amount of glucose in the blood

blood transfusion when blood from one person is put into another person (the two people must have compatible blood groups)

body mass index (BMI) measure of someone's weight in relation to their height

bone marrow centre of long bones

C

cancer life-threatening condition where body cells divide uncontrollably

capillaries small blood vessels that join arteries to veins

captive breeding breeding a species in zoos to maintain the wild population

carbohydrates chemicals found in all living things. They contain the elements carbon, hydrogen and oxygen. Sugars and starches are carbohydrates

carbon a very important element, carbon is present in all living things and forms a huge range of compounds with other elements

carbon cycle a natural cycle through which carbon moves by respiration, photosynthesis and combustion in the form of carbon dioxide

carbon dioxide (CO_2) gas present in the atmosphere at a low percentage but important in respiration, photosynthesis and combustion; a greenhouse gas which is emitted into the atmosphere as a by-product of combustion

carbon footprint the total amount of greenhouse gases given off by a person in a given time

carbon monoxide poisonous gas made when fuels burn in a shortage of oxygen

carnivore animal that eats other animals

carotene plant pigment involved in photosynthesis

cartilage softer and more flexible than bone, found in internal skeletons

cell differentiation when cells become specialised

cell membrane layer around a cell which helps to control substances entering and leaving the cell

central nervous system (CNS) collectively the brain and spinal cord

CFCs gases which used to be used in refrigerators and which harm the ozone layer

chemical digestion enzymes carry out chemical digestion when they break down large food molecules into smaller ones

chlorophyll pigment found in plants which is used in photosynthesis (gives green plants their colour)

cholesterol fatty substance which can block blood vessels

chromosomes thread-like structures in the cell nucleus that carry genetic information

clone genetically identical copy

closed circulatory system blood carried in blood vessels

coalesce join together/move as one

coherent waves having the same frequency, amplitude and phase (or constant phase difference)

collagen protein used for support in animal cells

colloids a liquid that has small particles dispersed it that do not settle

community all the plants and animals living in an ecosystem, e.g. a garden

compost dead and decaying plant material

compound fracture where a broken bone pierces skin

conservation a way of protecting a species or environment

contractile vacuole structure in amoeba used to remove excess water

coronary artery blood vessel that supplies the heart

corpuscles light particles as suggested by Newton in his particle theory of light

crenation when red blood cells shrink in concentrated solutions, they look partly deflated

crop rotation system of growing crops in sequence

D

decay to rot

decomposer organisms that break down dead animals and plants

defibrillator machine which gives the heart an electric shock to start it beating regularly

dehydration result of body losing too much water

denatured an enzyme is denatured if its shape changes so that the substrate cannot fit into the active site

denitrifying bacteria bacteria that convert nitrates into nitrogen gas

depressant a drug that slows down the working of the brain

dermis layer of skin

detritivores organisms that feed on decaying matter

detritus the dead and semi-decayed remains of living things

diastolic pressure the lowest point that your blood pressure reaches as the heart relaxes between beats

diet what a person eats

diffuse when particles diffuse they spread out

digestive system the metre-long system that handles and digests food (starts at mouth, ends at anus)

diploid cells that have two copies of each chromosome

DNA molecule found in all body cells in the nucleus – its sequence determines how our bodies are made (e.g. whether we have straight or curly hair), and gives each one of us a unique genetic code

DNA fingerprint identification obtained by examining a person's unique sequence of DNA

dominant allele/characteristic an allele that will produce the characteristic if present

double circulatory system blood system of two circuits, found in mammals

E

EAR for protein estimated average daily requirement of protein in diet

ecological niche the role of an organism within an ecosystem

egestion expulsion of solid waste

endangered a species where the numbers are so low they could soon become extinct

endocrine system the body system that is made up of endocrine glands (which secrete hormones)

energy the ability to 'do work', for example the human body needs energy to function

enzymes biological catalysts that increase the speed of a chemical reaction

epidermis the outer layer of the skin

essential elements the three elements, nitrogen, phosphorus and potassium that are essential for the growth of plants

eutrophication when waterways become too rich with nutrients (from fertilisers) which allows algae to grow wildly and use up all the oxygen

evaporation when a liquid changes to a gas, it evaporates

evolution the gradual change in organisms over millions of years caused by random mutations and natural selection

excretion the process of getting rid of waste from the body

exponential growth the ever-increasing growth of the human population

external skeleton skeleton on outside of body, usually of chitin

extinct when all members of a species have died out

F

fermentation the process of using yeast to break down sugars to alcohol

fertilisation when a sperm fuses (joins with) an egg

fertiliser chemical put on soil to increase soil fertility and allow better growth of crop plants

first-class protein proteins from meat and fish which contain all essential amino acids

flaccid floppy

follicle a ball of cells in the ovary containing an 'unripened' egg cell

follicle-stimulating hormone (FSH) the hormone in human females (made in the pituitary) which stimulates a follicle in an ovary to develop into a mature egg

fossil fuels fuels such as coal, oil and gas

fracture break in bone

fungi living organisms which can break down complex organic substances (some are pathogens and harm the body)

fungicide chemical used to kill fungi

Glossary

G

gametes the male and female sex cells (sperm and eggs)

gaseous exchange the movement of gases across an exchange membrane, e.g. in the lungs of mammals – gaseous exchange usually involves carbon dioxide and oxygen moving in opposite directions

gasohol biofuel that contains alcohol and petrol

gene section of DNA that codes for a particular characteristic

gene pool the different genes available within a species

gene therapy medical procedure where a virus is used to 'carry' a gene into the nucleus of a cell (this is a new treatment for genetic diseases)

genetic engineering transfer of genes from one organism to another

genotype the genetic makeup of an organism

geotropism a plant's growth response to gravity

global warming the increase in the Earth's temperature due to increases in carbon dioxide levels

glomerulus a tiny ball-shaped clump of blood vessels that filters substances from blood (found in kidney tubules)

glycerol together with fatty acids, these make up fats

gravity an attractive force between objects (dependent on their mass)

greenhouse gas any of the gases whose absorption of infrared radiation from the Earth's surface is responsible for the greenhouse effect, e.g. carbon dioxide, methane, water vapour

growth hormone a hormone produced by the pituitary gland that stimulates growth

H

habitat where an organism lives, e.g. the worm's habitat is the soil

haemoglobin chemical found in red blood cells which carries oxygen

hallucinogen a drug, like LSD, that gives the user hallucinations

haploid cells that have only one copy of each chromosome

heart attack damage to heart muscle, can be fatal

heat stroke result of body being too hot; skin is cold, pulse is weak

herbicide chemical used to kill weeds

heterozygous a person who has two different alleles for an inherited characteristic, e.g. someone with blond hair may also carry an allele for red hair

homozygous a person who has two alleles that are the same for an inherited feature, e.g. a blue-eyed person will have two blue alleles for eye colour

hormones chemicals that act on target organs in the body (hormones are made by the body in special glands)

humus organic matter in soil

hybrid the infertile offspring produced when two animals of different species breed

hydroponics growing plants in mineral solutions without the need for soil

hypothalamus small gland in brain, detects temperature of blood

hypothermia a condition caused by the body getting too cold, which can lead to death if untreated

I

immobilised enzyme enzyme inside gel beads

inbreeding breeding closely related animals

incubation period time between infection by a pathogen and when the first symptoms appear

indicator species organisms used to measure the level of pollution in water or the air

indicators chemicals which change colour according to the pH (indicators show how acid or alkali a substance is)

insecticide a chemical that can kill an insect

insulin hormone made by the pancreas which controls the level of glucose in the blood

internal skeleton skeleton inside body, made of cartilage or bone

K

kite diagram method of displaying results from a transect line

kwashiorkor an illness caused by protein deficiency due to lack of food. Sufferers often have swollen bellies caused by retention of fluid in the abdomen

L

ligament tissue joining bone to bone

ligase enzyme used to stick DNA together

lipase enzyme that break down fats into fatty acids and glycerol

luteinising hormone (LH) hormone (made in the pituitary) which, together with FSH, controls the release of an egg from the ovary

lysis to split apart

M

mammal animals that have hair or fur and produce milk for their young

marine snow organic matter that falls from the surface of oceans to the depths where it is used for food

meiosis cell division that results in haploid cells

menstrual cycle monthly hormonal cycle that starts at puberty in human females

meristem tips of roots and shoots where cell division and elongation takes place

metabolic rate the speed at which the amount of energy the body needs is released

microbes tiny microscopic organisms

microorganism very small organism (living thing) which can only be viewed through a microscope – also known as a microbe

microvilli microscopic projections from cells lining small intestine

minerals natural solid materials with a fixed chemical composition and structure, rocks are made of collections of minerals; mineral nutrients in our diet are things like calcium and iron, they are simple chemicals needed for health

mitochondria structures in a cell where respiration takes place

mitosis cell division that results in genetically identical diploid cells

molecule two or more atoms which have been chemically combined

motor neurone nerve cell carrying information from the central nervous system to muscles

Glossary

multicellular organism organisms made up of many specialised cells

mutation where the DNA within cells have been altered (this happens in cancer)

mutualism relationship in which both organisms benefit

N

natural selection process by which 'good' characteristics that can be passed on in genes become more common in a population over many generations ('good' characteristics mean that the organism has an advantage which makes it more likely to survive)

nitrogen-fixing bacteria bacteria that convert nitrogen into ammonia or nitrates

non-renewable energy energy which is used up at a faster rate than it can be replaced e.g. fossil fuels

nucleus central part of an atom that contains protons and neutrons

O

obesity a medical condition where the amount of body fat is so great that it harms health

oestrogen female hormone secreted by the ovary and involved in the menstrual cycle

open circulatory system blood system where blood is not contained in blood vessels

optimum conditions the conditions under which a reaction works most effectively

optimum temperature the temperature range that produces the best reaction rate

osmosis type of diffusion: movement of water from an area of high water concentration to an area of low water concentration

ossification the formation of bone from cartilage

osteoporosis disease of the bones in which bones become very brittle

ozone layer layer of the Earth's atmosphere that protects us from ultraviolet rays

P

pacemaker cells in heart that generate nerve impulses to stimulate muscle contraction

painkiller a drug that stops nerve impulses so pain is not felt

palisade cells tightly packed together cells found on the upper side of a leaf

parasite organism which lives on (or inside) the body of another organism

partially permeable membrane a membrane that allows some small molecules to pass through but not larger molecules

particulates particles such as soot released when fossil fuels burn

pasteurisation heating of a liquid to about 78 °C for several minutes to kill microorganisms

pathogen harmful organism which invades the body and causes disease

performance enhancer a drug used to improve performance in a sporting event

pesticide residue unwanted residues sometimes found in water contaminated by local pesticide use

petrol volatile mixture of hydrocarbons used as a fuel

pH scale scale in which acids have a pH below 7, alkalis a pH of above 7 and a neutral substance a pH of 7

phloem specialised transporting cells which form tubules in plants to carry sugars from leaves to other parts of the plant

photosynthesis process carried out by green plants where sunlight, carbon dioxide and water are used to produce glucose and oxygen

phototropism a plant's growth response to light

physical digestion breaking down of food particles by teeth or muscles

phytoplankton microscopic organisms in water that can photosynthesise

plant hormones hormones that control various plant processes such as growth and germination

plaque build up of cholesterol in a blood vessel (which may block it)

plasma yellow liquid found in blood

plasmid circular DNA in bacteria

plasmolysis the shrinking of a plant cell due to loss of water; the cell membrane pulls away

platelets cell fragments which help in blood clotting

pollination the process of transferring pollen from one plant to another

pollute contaminate or destroy the environment

pollution contaminating or destroying the environment as a result of human activities

population group of organisms of the same species in a habitat

predator animal which preys on (and eats) another animal

prey animals which are eaten by a predator

producers organisms in a food chain that make food using sunlight

progesterone hormone produced by the ovary, which prepares the uterus for pregnancy

protease enzyme which break down proteins into amino acids

R

radioactive waste waste produced by radioactive materials used at nuclear power stations, research centres and some hospitals

radiotherapy using ionising radiation to kill cancer cells in the body

random having no regular pattern

recessive allele/characteristic two recessive alleles needed to produce the characteristic

recycle to reuse materials

red blood cells blood cells which are adapted to carry oxygen

reflex a muscular action that we take without thinking about

renewable energy energy that can be replenished at the same rate that it's used up, e.g. biofuels

reptile cold blooded vertebrate having an external covering of scales or horny plates

respiration process occurring in living things where oxygen is used to release the energy in foods

restriction enzyme enzyme used to cut DNA

rhesus blood can be grouped into rhesus-positive and rhesus-negative groups

ribosome structures in a cell where protein synthesis takes place

S

saprophyte an organism that breaks down dead organic matter, usually used to refer to fungi

saturated fat fats, most often of animal origin, which are solid at room temperature

second-class protein proteins from plants which only contain some essential amino acids

selective breeding process of breeding organisms with the desired characteristics

sensory neurone nerve cell carrying information from receptors to central nervous system

sex chromosomes a pair of chromosomes that determine gender, XX in female, XY in male

simple fracture clean break in bone

single circulatory system blood system with only one circuit, e.g. in fish

soluble a soluble substance can dissolve in a liquid, e.g. sugar is soluble in water

species basic category of biological classification, composed of individuals that resemble one another, can breed among themselves, but cannot breed with members of another species

spongy mesophyll cells found in the middle of a leaf with an irregular shape and large air spaces between them

stem cells unspecialised body cells (found in bone marrow) that can develop into other, specialised, cells that the body needs, e.g. blood cells

stimulant a drug that speeds up the working of the brain

stomata small holes in the surface of leaves which allow gases in and out of leaves

stroke sudden change in blood flow to the brain – can be fatal

sucrase enzyme that breaks down sucrose (sugar)

sustainable development managing a resource so that it does not run out

sweat liquid produced by the skin; it cools you down when it evaporates

synapse gap between two neurons

synovial joint joint containing synovial fluid

systolic pressure the highest point that your blood pressure reaches as the heart beats to pump blood through your body

T

tendon tissue joining muscle to bone

therapy treatment of a medical problem

thrombosis blood clot in a blood vessel causing it to be blocked

thyroid gland gland at the base of the neck which makes the hormone thyroxin

tissue culture process that uses small sections of tissue to clone plants

toxic a toxic substance is one which is poisonous, e.g. toxic waste

toxin a poisonous substance

transect line across an area to sample organisms

transgenic organism organism that contains DNA from another organism

trials tests to find if something works and is safe

trophic level the stages in a food chain

tumour abnormal mass of tissue that is often cancerous

turgid plant cells which are full of water with their walls bowed out and pushing against neighbouring cells

turgor pressure the pressure exerted on the cell membrane by the cell wall when the cell is fully inflated

U

units of alcohol measurement of alcoholic content of a drink

urea nitrogen-containing substance cleared from the blood by kidneys and excreted in urine

uric acid excretory product found in urine

V

variation the differences between individuals (because we all have slight variations in our genes)

vascular bundle group of xylem and phloem cells

vasoconstriction in cold conditions the diameter of small blood vessels near the surface of the body decreases – this reduces the flow of blood

vasodilation in hot conditions the diameter of small blood vessels near the surface of the body increases – this increases the flow of blood

vector an animal that carries a pathogen without suffering from it

vegan a type of diet; a person who does not eat animals or animal products

vegetarian a type of diet; a person who does not eat meat or fish

veins blood vessels that carry blood back to the heart

villi 'finger-like' structures on the surface of the small intestine which give it a greater surface area for absorption

viruses very small infectious organisms that reproduce within the cells of living organisms and often cause disease

W

white blood cells blood cells which defend against disease

X

xanthophylls plant pigments involved in photosynthesis

xylem cells specialised for transporting water through a plant; xylem cells have thick walls, no cytoplasm and are dead, their end walls break down and they form a continuous tube

Y

yeast single-celled fungus used in making beer and bread

Collins

Revision

New GCSE Biology

Exam Practice Workbook

Higher

For OCR Gateway B

Exam tips

The key to successful revision is finding the method that suits you best. There is no right or wrong way to do it.

Before you begin, it is important to plan your revision carefully. If you have allocated enough time in advance, you can walk into the exam with confidence, knowing that you are fully prepared.

Start well before the date of the exam, not the day before!

It is worth preparing a revision timetable and trying to stick to it. Use it during the lead up to the exams and between each exam. Make sure you plan some time off too.

Different people revise in different ways and you will soon discover what works best for you.

Remember

There is a difference between *learning* and *revising*.

When you revise, you are looking again at something you have already learned. Revising is a process that helps you to remember this information more clearly.

Learning is about finding out and understanding new information.

Using the Workbook

This Workbook allows you to work at your own pace and check your answers using the detachable Answer section on pages 137–144. In addition to the exam practice questions, the Workbook also contains questions that require longer answers (Extended response questions). You will find one question that is similar to these in each section of your written exam papers. The model answers supplied for these questions give guidance about the content that should be included, but do not necessarily provide a complete response for the questions concerned.

Some general points to think about when revising

- Find a quiet and comfortable space at home where you won't be disturbed. You will find you achieve more if the room is ventilated and has plenty of light.
- Take regular breaks. Some evidence suggests that revision is most effective when tackled in 30 to 40 minute slots. If you get bogged down at any point, take a break and go back to it later when you are feeling fresh. Try not to revise when you're feeling tired. If you do feel tired, take a break.
- Use your school notes, textbook and this Revision guide.
- Spend some time working through past papers to familiarise yourself with the exam format.
- Produce your own **summaries** of each module and then look at the summaries in this Revision guide at the end of each module.
- Draw mind maps covering the key information on each topic or module.
- Review the **Grade booster checklists** on pages 128–133.
- Set up revision cards containing condensed versions of your notes.
- Prioritise your revision of topics. You may want to leave more time to revise the topics you find most difficult.

Fitness and health

D–C

1 Joe's mum has her blood pressure checked by a nurse. The nurse tells her the results and she fills in a questionnaire for the nurse.

Blood pressure questionnaire

Questions	Notes	Answers Yes	Answers No
1 Do you take regular exercise?	Strong heart muscles will lower blood pressure		✓
2 Do you eat a healthy balanced diet?	Reducing salt intake will lower blood pressure		✓
3 Are you overweight?	Being overweight by 5 kg raises blood pressure by 5 units	✓	
4 Do you regularly drink alcohol?	A high alcohol intake will damage liver and kidneys	✓	
5 Are you under stress?	Relaxation will lower blood pressure	✓	

a Suggest **two** changes Joe's mum should make to lower her blood pressure.

.... **[2 marks]**

B–A*

b Explain why Joe's mum should lower her blood pressure.

.... **[2 marks]**

D–C

2 Look at the data about the risk factors linked to heart disease (USA data, 1961–2004).

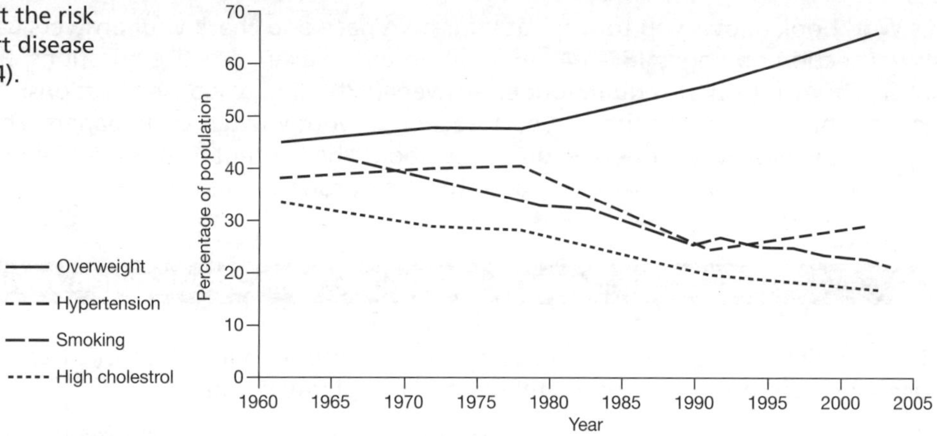

a Describe the general trends shown by this data.

.... **[2 marks]**

b What trend would you have expected to find in data about the **actual** heart disease in the same time period?

.... **[2 marks]**

D–C

3 Joe is an athlete and he is very fit. However, he still catches a cold. Explain why being fit may not keep you healthy.

.... **[2 marks]**

B–A*

4 Gas boilers in houses must be regularly checked to make sure they are not releasing carbon monoxide. Explain the effects of carbon monoxide on the body.

.... **[4 marks]**

Human health and diet

1 Children in developing countries sometimes have a swollen abdomen, a condition caused by a low-protein diet.

a Write down the name of this condition.

.... [1 mark]

D–C

b Describe two possible reasons for the lack of protein in developing countries.

....

.... [2 marks]

2 a Simon has a mass of 40 000 g. Calculate his estimated average daily requirement (EAR) for protein in grams. Use this formula: EAR in g = 0.6 × body mass in kg.

Show your working.

EAR = ... g [2 marks]

D–C

b Simon has a sister called Karen. He is worried that she is not eating enough food.

Suggest why Karen may have chosen not to eat enough food to meet her daily requirements.

....

....

.... [2 marks]

c Simon is a vegetarian. Although he eats enough protein every day to meet his EAR, he needs to think carefully about his protein intake.

Explain why.

....

....

....

.... [2 marks]

B–A*

3 Glycogen is a large molecule made up of hundreds of glucose molecules joined together.

glucose units

glycogen is a complex carbohydrate

a What group of food chemicals does glycogen belong to?

.... [1 mark]

D–C

b Proteins are also made of hundreds of small molecules that are joined together in a chain.

Write down the name of these molecules.

.... [1 mark]

c Sometimes a person takes in more glycogen and protein than they need.

Compare how the body deals with the extra glycogen and protein that is taken in.

....

.... [2 marks]

B–A*

Staying healthy

1 Complete the sentences about infectious diseases. Use words from this list.

antibodies **antibiotics** **antigens** **hormones** **pathogens** **toxins** **vectors**

a The symptoms of an infectious disease are caused by ....

They produce chemical waste, which contains ....

b Each disease-causing microorganism has its own .... so

the body needs specific .... to combat them. **[2 marks]**

2 Look at the diagram. It shows how mosquitoes spread malaria.

a Using the diagram, name a parasite and its host.

....

.... **[1 mark]**

b Use the diagram to explain **two** ways in which the spread of malaria could be controlled.

....

....

....

....

.... **[4 marks]**

3 The graph shows the levels of immunity to a disease using passive and active immunity methods.

a Describe the difference in immunity:

i at 20 days .... **[1 mark]**

ii after 60 days .... **[1 mark]**

b Explain why there is a difference in immunity between passive and active immunity.

....

....

.... **[4 marks]**

c Which method of immunity should be used in an outbreak of infectious disease following a natural disaster such as a tsunami? Explain your choice.

.... **[1 mark]**

d Using MRSA as an example, explain why doctors are concerned about the overuse of antibiotics.

....

....

.... **[3 marks]**

The nervous system

1 a Rashid does his homework on how the eye works. This is what he writes. There are **three** mistakes.

The light rays are reflected by the cornea and lens. An image is formed on the optic nerve. Nerve impulses are then sent to the spinal cord. The amount of light entering the eye is controlled by the iris.

Write down the incorrect word(s) and their correct replacements.

.... should be ....;

.... should be ....;

.... should be .... **[3 marks]**

D–C

b Write down **two** ways in which vision will be affected if only one eye is used to look at an object.

.... **[2 marks]**

c Use the information in the diagram to explain how the eye accommodates.

....

....

....

.... **[4 marks]**

B–A*

2 a Look at the diagram of a motor neurone.

Write down the names of the parts A, B, and C.

A ....

B ....

C ....

[3 marks]

b Which part A, B or C, carries the nerve impulse? .... **[1 mark]**

D–C

c Look at the sequences A, B, C and D in a reflex arc. Only one sequence is correct.

A effector⇒sensory neurone⇒motor neurone⇒central nervous system⇒response

B stimulus⇒receptor⇒sensory neurone⇒motor neurone⇒central nervous system⇒response

C stimulus⇒receptor⇒sensory neurone⇒central nervous system⇒motor neurone⇒response

D response⇒receptor⇒sensory neurone⇒central nervous system⇒motor neurone⇒stimulus

Which sequence is correct? .... **[1 mark]**

d Explain how a nerve impulse is transmitted from one neurone to the next.

....

....

.... **[5 marks]**

B–A*

Drugs and you

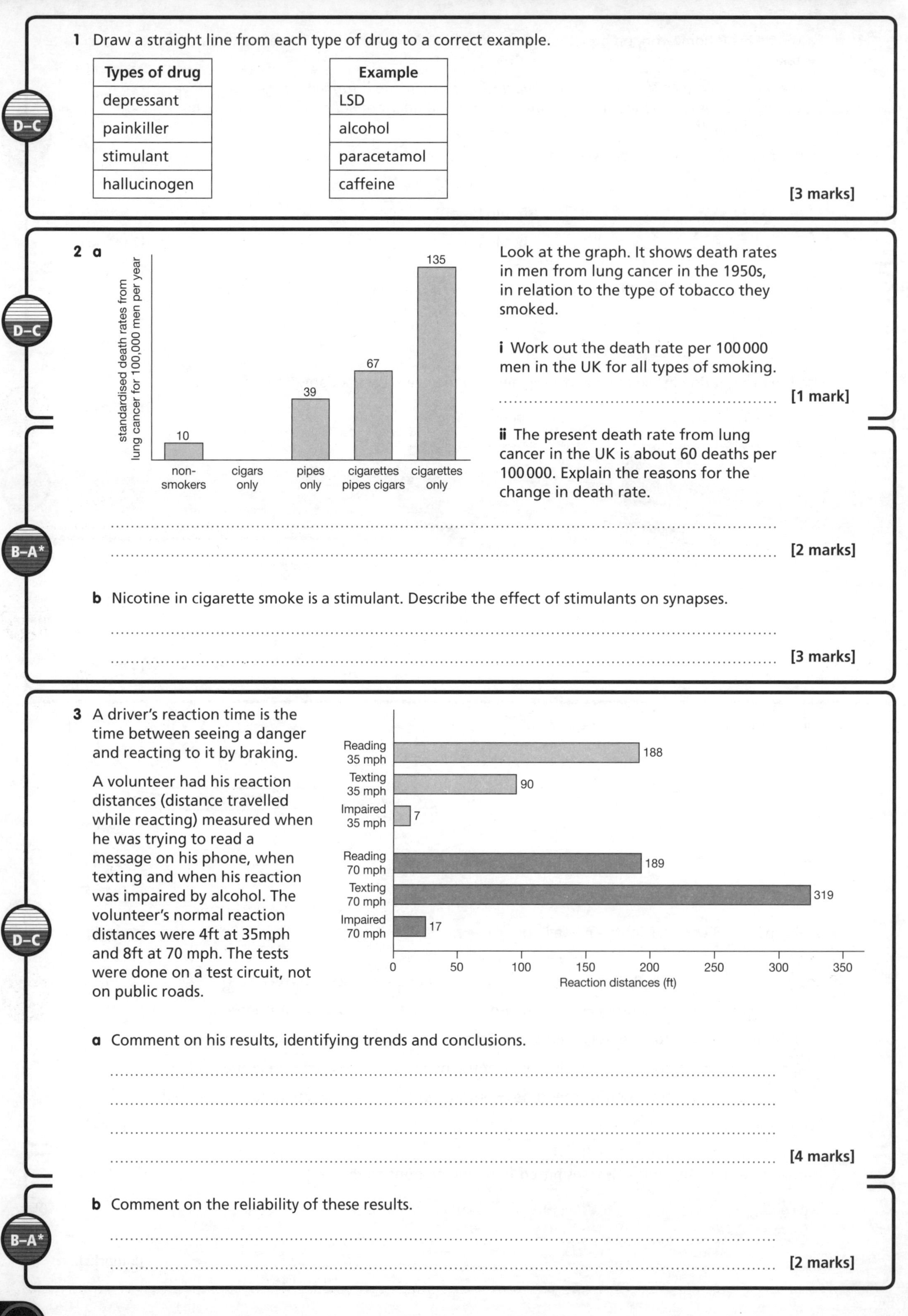

D–C

1 Draw a straight line from each type of drug to a correct example.

Types of drug
depressant
painkiller
stimulant
hallucinogen

Example
LSD
alcohol
paracetamol
caffeine

[3 marks]

D–C

2 a Look at the graph. It shows death rates in men from lung cancer in the 1950s, in relation to the type of tobacco they smoked.

i Work out the death rate per 100 000 men in the UK for all types of smoking.

.... **[1 mark]**

B–A*

ii The present death rate from lung cancer in the UK is about 60 deaths per 100 000. Explain the reasons for the change in death rate.

....

.... **[2 marks]**

b Nicotine in cigarette smoke is a stimulant. Describe the effect of stimulants on synapses.

....

.... **[3 marks]**

D–C

3 A driver's reaction time is the time between seeing a danger and reacting to it by braking.

A volunteer had his reaction distances (distance travelled while reacting) measured when he was trying to read a message on his phone, when texting and when his reaction was impaired by alcohol. The volunteer's normal reaction distances were 4ft at 35mph and 8ft at 70 mph. The tests were done on a test circuit, not on public roads.

a Comment on his results, identifying trends and conclusions.

....

....

....

.... **[4 marks]**

B–A*

b Comment on the reliability of these results.

....

.... **[2 marks]**

Staying in balance

1 a What is meant by homeostasis?

.... **[1 mark]**

b The body can increase or decrease heat transfer to the outside environment. What is the importance of the body temperature being 37 °C?

.... **[1 mark]**

D–C

c If the body gets too hot you can suffer from dehydration. Explain why.

....

.... **[2 marks]**

d The temperature of an incubator for premature babies is controlled by a negative feedback mechanism. If the temperature gets too low, the heater is switched on.

i Explain what is meant by negative feedback.

.... **[2 marks]**

B–A*

ii Explain how negative feedback mechanisms are used in homeostasis.

....

.... **[3 marks]**

2 a Type 1 diabetes is treated by hormone injections.

Explain why Type 2 diabetes can be controlled by diet.

....

....

.... **[3 marks]**

D–C

b Look at the graph showing the blood sugar levels of two people after drinking a glucose solution.

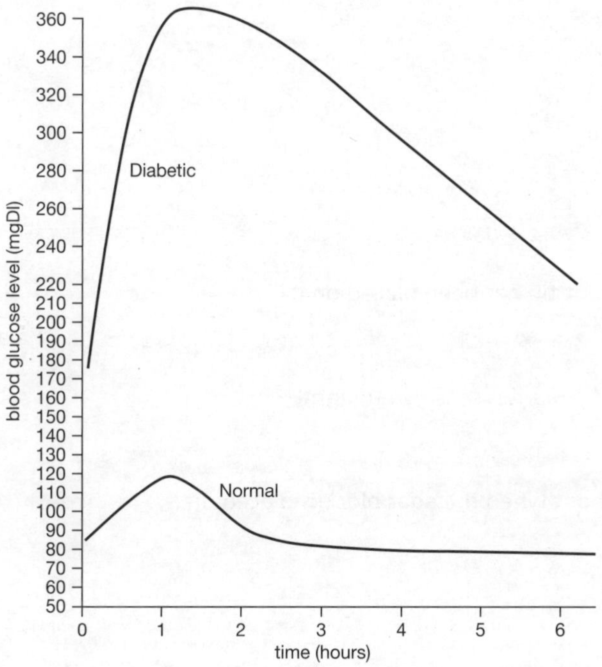

blood glucose level (mgDl)
360
340
320
300
280
260
240
220
210
200
190
180
170
160
150
140
130
120
110
100
90
80
70
60
50
Diabetic
Normal
0
1
2
3
4
5
6
time (hours)

i Why did the blood glucose levels rise in the first hour?

....

....

.... **[1 mark]**

ii Explain why the blood glucose levels for the 'normal' person returned to its original level after two hours.

....

....

....

....

....

....

.... **[3 marks]**

B–A*

iii Explain why the blood glucose levels remained high in the diabetic person.

....

.... **[2 marks]**

Controlling plant growth

D–C

1 Charlotte grows fruit trees.

a She takes cuttings of the best trees and dips the ends in rooting powder.

Describe and explain the effect the rooting powder has on the cuttings.

....

.... **[2 marks]**

b Charlotte has many weeds in her lawn. She uses a selective weedkiller.

i Explain what is meant by a selective weedkiller.

....

.... **[2 marks]**

ii How does a selective weedkiller work?

....

.... **[2 marks]**

D–C

2 a Complete the sentences about tropisms. Use words from this list.

auxins antigens geotropic gravity heat light negatively positively

Since shoots grow towards the light they are called .... phototropic.

Since roots grow with the pull of .... they are called

positively ....

These reactions involve plant hormones called .... **[4 marks]**

B–A*

b Look at the diagram. It shows the result of an experiment on a shoot tip.

block

agar block placed on cut shoot

i What will the agar block contain after the shoot tip has been placed on it?

.... **[1 mark]**

ii Using this information, suggest which part of a plant is sensitive to light.

.... **[1 mark]**

iii Describe and explain what happens to the shoot when the agar block is placed on it.

....

....

....

....

....

.... **[6 marks]**

Variation and inheritance

1 Mutations can cause genetic variation in organisms.

a What is a mutation?

.... **[1 mark]**

D–C

b Explain **two** other causes of genetic variation.

1 ....

2 .... **[4 marks]**

2 Complete the sentences about chromosomes.

Use words from this list.

12 23 46 the same a different a random

Most human body cells have .... number of chromosomes.

Most human body cells have .... pairs of chromosomes.

Other species have .... number of chromosomes. **[4 marks]**

D–C

3 a Look at the diagram showing how gender is inherited.

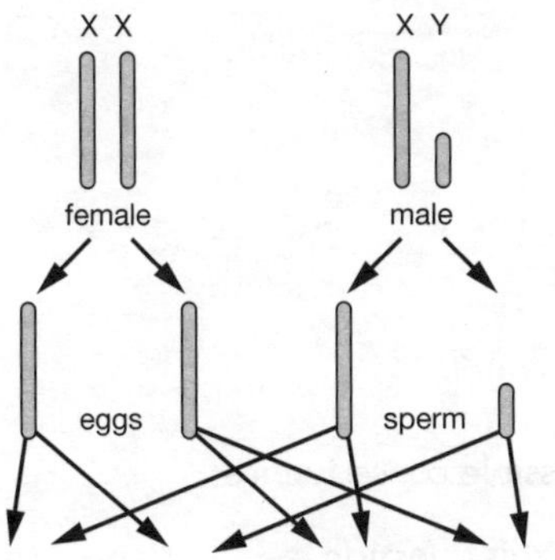

X X

female

X Y

male

eggs

sperm

Complete the diagram by writing a letter at the end of each of the eight arrows at the bottom, to show the sex chromosomes and gender of the possible genetic combinations. **[3 marks]**

B–A*

b Use the diagram to explain why there is an equal chance of a baby being male or female.

....

....

....

.... **[3 marks]**

c Cystic fibrosis is an inherited condition. Darren has cystic fibrosis but his parents have not.

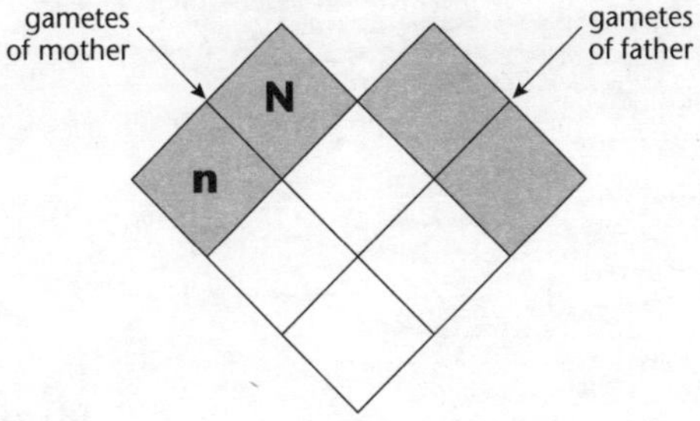

gametes of mother

gametes of father

N

n

i Complete the genetic diagram to show how Darren inherited cystic fibrosis. **[2 marks]**

ii Put a ring around Darren's genotype in the diagram. **[1 mark]**

B1 Extended response question

Blood pressure

Four adult friends measure their blood pressure. They then look at a blood pressure chart.

	Blood pressure in mmHg	
	Systolic	Diastolic
Alan	160	95
Rick	110	70
Shabeena	130	95
Toni	100	85

In the blood pressure chart, only one of the numbers needs to be higher/lower than it should be to count as either high or low blood pressure.

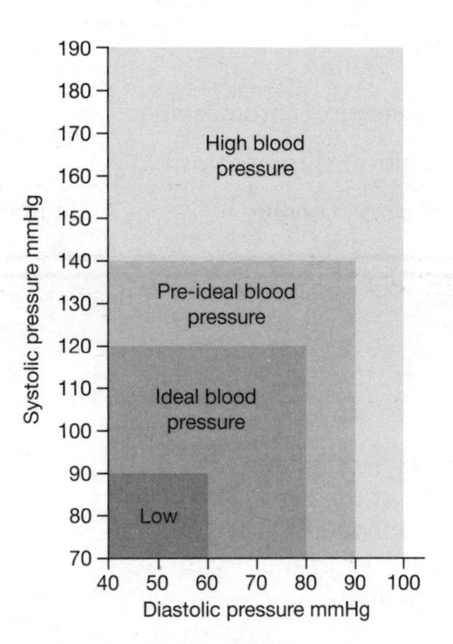

Suggest reasons for the friends' blood pressure readings and explain the possible consequences.

The quality of written communication will be assessed in your answer to this question.

....

....

....

....

....

....

....

....

....

....

....

....

....

....

....

....

.... **[6 marks]**

Classification

1 a This table shows the classification of lions. Complete it by inserting the missing groups.

kingdom	Animal
phylum	Chordate
class	Mammal
order	Carnivore
family	Felidae
genus	Panthera
species	leo

[5 marks]

D–C

b There are seven different types of lion alive today. Describe one advance in science that could be used to help classify them that would not have been available 60 years ago.

Natural system and artificial system could be used to help classify them.

[2 marks]

B–A*

2 a What is meant by the term species?

A group of organisms that can interbreed to produce fertile offsprings. **[2 marks]**

b Lions and tigers belong to the same family of cats. Look at the table. It shows the Latin names of some different cats.

Common name	Latin name
bobcat	*Felix rufus*
cheetah	*Acinonyx jubatus*
lion	*Panthera leo*
ocelot	*Felix pardalis*

i Two of the cats are more closely related. Write down the common names of these two cats.

Bobcat and an ocelot

[1 mark]

ii Explain your answer to part (bi).

Their first latin name is Felix.

[1 mark]

D–C

3 Scientists discovered a fossil of an extinct animal called *Archaeopteryx*. It had some bird features and some reptile features. Explain why there are always going to be some organisms like this that are difficult to classify.

Two different species interbreed to produce an unfertile offspring. They're difficult to classify because ~~they~~ has characteristics from a bird and an reptile.

[2 marks]

D–C

4 A zorse is a cross between two different species: a zebra and a horse.

a What name is used to describe a cross between two different species?

Hybrid

[1 mark]

b The zorse is difficult to classify. Explain why.

~~it's~~ It had characteristics from a zebra and a horse

[1 mark]

B–A*

5 Camels live in deserts in Asia. Llamas live in dry areas in South America. An animal similar to the llama and camel lived in North America about 11 million years ago. The appearance of a camel and a llama has a number of similarities. Explain possible reasons why.

Both the camels and a llama live in the same habitat.

[3 marks]

B–A*

Energy flow

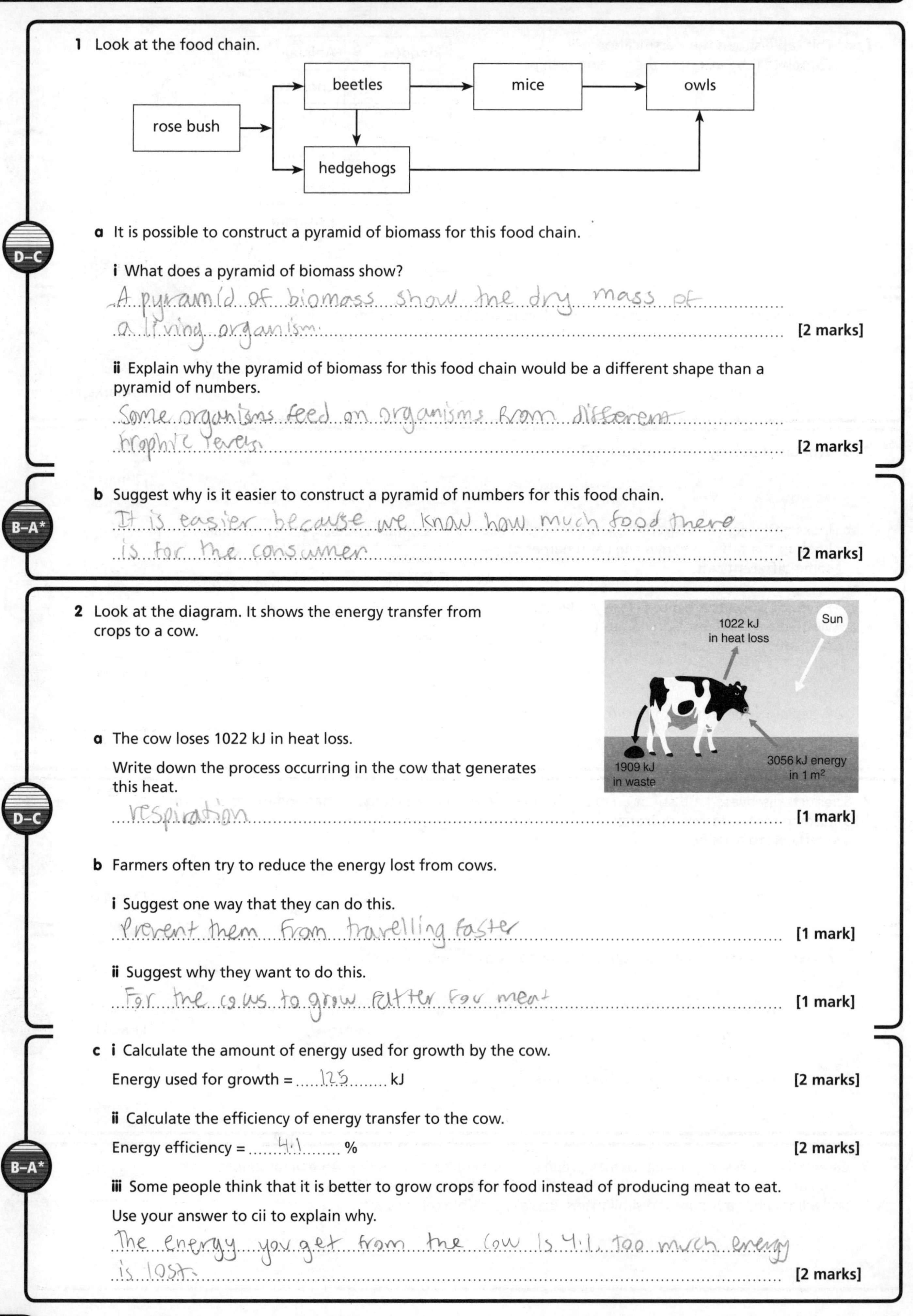

1 Look at the food chain.

a It is possible to construct a pyramid of biomass for this food chain.

i What does a pyramid of biomass show?

A pyramid of biomass show the dry mass of a living organism. **[2 marks]**

ii Explain why the pyramid of biomass for this food chain would be a different shape than a pyramid of numbers.

Some organisms feed on organisms from different trophic levels **[2 marks]**

b Suggest why is it easier to construct a pyramid of numbers for this food chain.

It is easier because we know how much food there is for the consumer **[2 marks]**

2 Look at the diagram. It shows the energy transfer from crops to a cow.

a The cow loses 1022 kJ in heat loss.

Write down the process occurring in the cow that generates this heat.

respiration **[1 mark]**

b Farmers often try to reduce the energy lost from cows.

i Suggest one way that they can do this.

Prevent them from travelling faster **[1 mark]**

ii Suggest why they want to do this.

For the cows to grow fatter for meat **[1 mark]**

c **i** Calculate the amount of energy used for growth by the cow.

Energy used for growth = 125 kJ **[2 marks]**

ii Calculate the efficiency of energy transfer to the cow.

Energy efficiency = 4.1 % **[2 marks]**

iii Some people think that it is better to grow crops for food instead of producing meat to eat.

Use your answer to cii to explain why.

The energy you get from the cow is 4.1. too much energy is lost. **[2 marks]**

Recycling

1 The diagram shows the carbon cycle.

a Finish labelling the cycle by writing the correct processes in the three blank labels. **[3 marks]**

b Decomposers return carbon dioxide to the air.

They will release carbon dioxide more slowly from waterlogged soils.

Explain why.

Decomposers cannot respire and there's less oxygen in waterlogged soils. **[2 marks]**

D–C

c Animals called corals live in many oceans. The carbon in coral is recycled over millions of years.

Explain how this occurs.

It turns into limestone and is being weathered by acid rain. **[3 marks]**

B–A*

2 a Why do plants need nitrates?

For growth **[1 mark]**

D–C

b The diagram shows part of the nitrogen cycle.

nitrogen in organic compounds in plant leaves → (decomposers) → compound X → (bacteria Y) → nitrates

Write down the name of:

i compound X ammonia **[1 mark]**

ii bacteria Y Nitrifying bacteria **[1 mark]**

B–A*

c Nitrogen gas in the air is unreactive.

Describe how it can be converted into a form that can be used by plants.

By using nitrifying bacteria. **[4 marks]**

Interdependence

1 Read the information about the red and grey squirrels.

> There are two main species of squirrel living in Britain, the red squirrel and the grey squirrel. The grey was introduced from America. The number of reds has gone down in Britain over the last 60 years. They are now rare. There have been a number of studies to try to find out why the number of reds is declining.
>
> In coniferous woodland the reds are lighter animals and so spend more time up in the trees, feeding on the seeds from pine cones.
>
> It is in deciduous woodland that the reds are disappearing. Both types try to feed on acorns on the floor of the forest but the greys can digest the acorns more easily.

D–C

a What do red squirrels compete with each other for that they do not compete with grey squirrels for?

Mates .... **[1 mark]**

B–A*

b i The red squirrel has a particular ecological niche. What is meant by 'ecological niche'?

Ecological niche is the habibat that an organism lives in. .... **[2 marks]**

ii Use the term ecological niche to explain why red and grey squirrels compete more in deciduous forests.

.... **[2 marks]**

2 The graph shows the predator–prey relationship of lemmings and snowy owls.

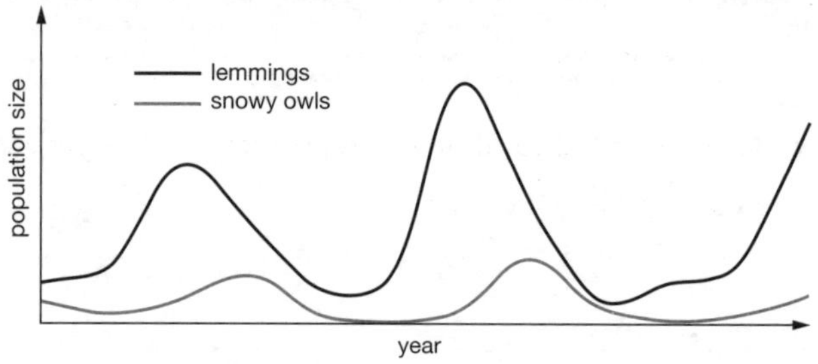

D–C

a Write about why the number of lemmings change, as shown in the graph.

....

.... **[2 marks]**

B–A*

b Why do the owl numbers peak a short time after the lemming numbers?

.... **[2 marks]**

D–C

3 Oxpeckers are birds that eat insects on buffaloes. Buffalo often suffer from insect parasites. The oxpecker and the buffalo both benefit from the relationship.

Write down the name given to this type of relationship.

.... **[1 mark]**

4 Pea plants contain bacteria in special nodules on their roots. Pea plants are pollinated by bees. These bees often have small animals called mites that feed on fluid from the bee's body.

D–C

a Write down an organism that is acting as a parasite in these feeding relationships.

.... **[1 mark]**

b The pea plant and the bee both gain from their relationship. Explain how.

.... **[2 marks]**

B–A*

c How do the bacteria and the pea plant both gain from their relationship?

....

.... **[2 marks]**

Adaptations

1 a Polar bears are adapted to live in the cold.

Suggest ways that the polar bear may be adapted to live in the cold.

....

....

.... **[3 marks]**

D–C

b The black bear and the polar bear both live in Canada. Both types of bears spend the winter months buried in holes or dens.

Explain why they do this.

....

....

.... **[3 marks]**

c The black bear lives further south than the polar bear. A female black bear has a mass of 40–80 kg, whereas a male polar bear is 150–250 kg. Black bears also have much larger ears than polar bears.

Explain how these differences are related to the different habitats of the bears.

B–A*

....

....

.... **[3 marks]**

2 a Cacti live in deserts.

The cactus has spines instead of leaves. Explain why.

....

.... **[2 marks]**

D–C

b Lizards called gila monsters live in the same habitat as cacti. They hunt small mammals.

In the morning they lie out in the Sun for some time before they hunt. Explain why.

....

.... **[2 marks]**

3 Raccoons are mammals about the size of small dogs that live throughout most of North America.

They are omnivores, eating a very wide range of foods.

Explain how this allows them to survive changing conditions and live over such a wide area.

B–A*

....

.... **[2 marks]**

Natural selection

1 Question 1 on page 86 includes a passage about competition between the red and grey squirrel. Re-read the passage before attempting these questions.

D–C

a What adaptation allows the grey squirrels to be better adapted to living in deciduous woodland?

.... **[1 mark]**

b Many scientists hope that a new strain of the red squirrel will evolve that can digest acorns. Use Darwin's theory of natural selection to explain how this might come about.

....

.... **[2 marks]**

B–A*

c Red squirrels now only live in isolated populations in different parts of the country. How might this lead to new species of squirrels developing?

....

....

....

.... **[4 marks]**

2 The following article gives information on the superbug MSRA. Read it carefully.

D–C

> **Where did it come from?**
>
> MRSA evolved because of natural selection. There are lots of different strains of the bacteria. Each strain has slightly different DNA. The DNA is also constantly mutating as the bacteria reproduce. Some of these mutations will be more resistant to antibiotics than others. When people take antibiotics, the less resistant strains die first. The more resistant strains are harder to destroy. If people stop taking the antibiotics too soon, the resistant strains survive.

Use Darwin's theory of natural selection to explain how MRSA has evolved.

....

.... **[2 marks]**

3 On his voyage on the *Beagle*, Charles Darwin visited many small islands. He made a number of observations that helped him to develop his theory of natural selection. One observation was that on small islands, animals often evolve to produce smaller animals than on the mainland.

D–C

a When Darwin returned from his voyage he was rather worried about publishing his ideas about natural selection. Suggest why that was.

....

.... **[2 marks]**

B–A*

b Lamarck's theory of evolution could also have explained why animals on an island are usually smaller. How would Lamarck's theory have explained this observation?

....

.... **[2 mark]**

c Why would this explanation be considered to be incorrect by most scientists now?

.... **[1 mark]**

Population and pollution

1 a The rise in human population is causing an increased level of carbon dioxide in the air. Suggest **two** effects this increase may have on the environment.

.... **[2 marks]**

b The ozone layer in the Earth's atmosphere protects us from harmful ultraviolet rays. Chemicals are destroying the ozone layer.

i Write down the name of the chemicals.

.... **[1 mark]**

ii The overuse of these chemicals has caused an increase in skin cancer. Suggest a reason why.

.... **[1 mark]**

D–C

2 Look at the graph. It shows the past, present and predicted future world human population.

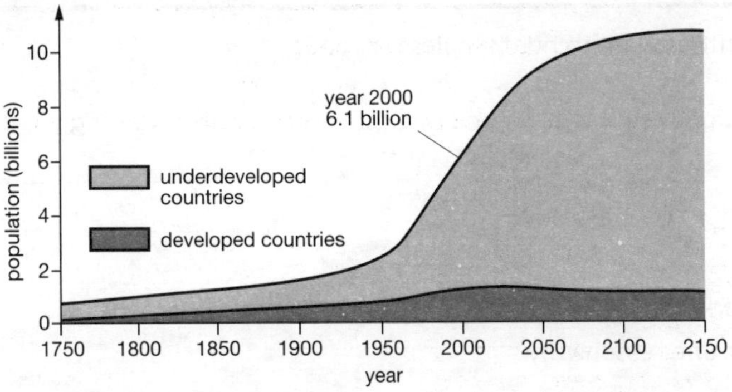

a Human population is in the rapid growth stage. Write down the name given to this stage of growth.

....

.... **[1 mark]**

D–C

b In developed countries such as America the population is constant.

In underdeveloped countries the population may be higher than in developed countries, yet they cause less pollution. Suggest two reasons why.

....

.... **[2 marks]**

B–A*

3 Scientists can look for the variety of animal species living in a stream when they want to measure how polluted the water is.

a Write down the name given to species that are used to measure levels of water pollution.

.... **[1 mark]**

b Look at the table. It shows the sensitivity of different animals to pollution.

A river sample contained mussels, damsel fly larvae and bloodworms, but no mayfly or stonefly larvae. Use this information to explain how you can tell that the river is polluted.

Animal	Sensitivity to pollution
stonefly larva	sensitive
water snipe fly	sensitive
alderfly	sensitive
mayfly larva	semi-sensitive
freshwater mussel	semi-sensitive
damselfly larva	semi-sensitive
bloodworm	tolerates pollution
rat-tailed maggot	tolerates pollution
sludgeworm	tolerates pollution

....

....

.... **[2 marks]**

D–C

c If a river is polluted, the water usually contains less oxygen.
Scientists could use oxygen probes to give an indication of the level of pollution.

Give **one** advantage and **one** disadvantage of this method rather than looking at the variety of animals in the river.

....

.... **[2 marks]**

B–A*

Sustainability

D–C

1 **a** Pandas live in a remote part of China. Their habitat is being destroyed. Some people want to save the panda from extinction.

Suggest **two** reasons why saving the panda might help the people who live in the same habitat.

....

.... **[2 marks]**

B–A*

b Pandas are being bred in zoos in China and in other countries. A careful record is kept about the family tree of each panda and this is consulted before they are allowed to mate.

Explain why this is.

....

.... **[2 marks]**

2 Some countries want to hunt whales for food.

D–C

a Suggest **one** argument for and **one** argument against hunting whales.

....

.... **[2 marks]**

B–A*

b It is very difficult to stop people hunting whales.

Suggest **one** reason why.

.... **[1 mark]**

3 The fishing industry is trying to follow sustainable development.

a What is meant by sustainable development?

....

.... **[2 marks]**

D–C

b To try and achieve this, the government has set fish quotas.
Fishermen can only catch a set amount of fish at any one time.
Also, the size of the individual fish that they catch has to be above a certain level.
These regulations should help maintain the population of fish in the sea.

Explain why.

....

.... **[2 marks]**

c The population of Brazil was estimated at 150 million in 1990 and 195 million in 2010.

Explain the problems of achieving sustainable development when the population is rising this rapidly.

B–A*

....

....

.... **[3 marks]**

The diagram shows the flow of energy through a food chain.

sunlight → **wheat** 10 000 units → **cows** 1000 units → **humans** 200 units

(wheat → 9000 units; cows → 800 units)

Use your knowledge of how energy is lost from food chains to explain the figures shown and discuss how the figures could be used to argue for vegetarianism.

The quality of written communication will be assessed in your answer to this question.

[6 marks]

Molecules of life

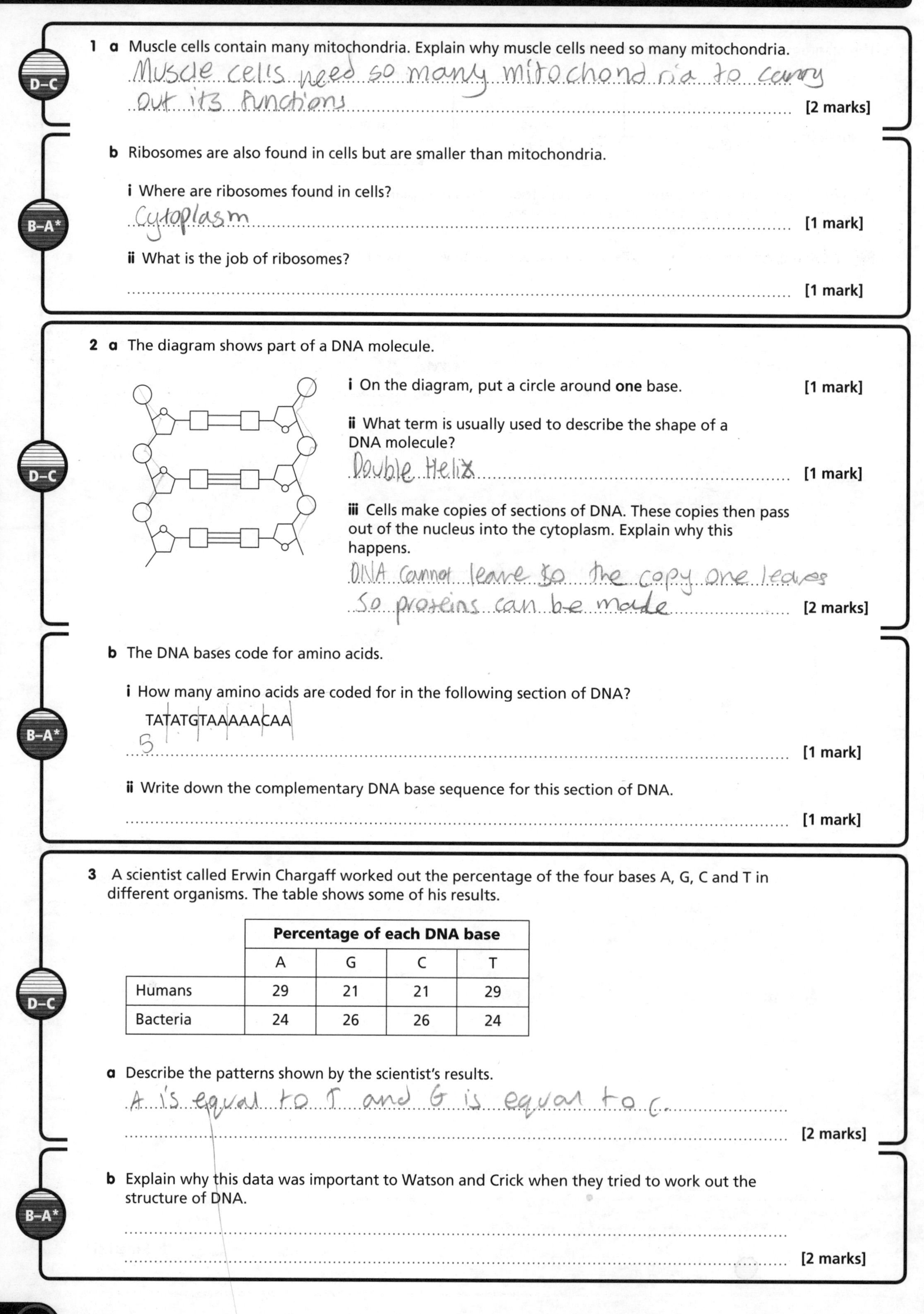

D–C

1 a Muscle cells contain many mitochondria. Explain why muscle cells need so many mitochondria.

Muscle cells need so many mitochondria to carry out its functions

[2 marks]

B–A*

b Ribosomes are also found in cells but are smaller than mitochondria.

i Where are ribosomes found in cells?

Cytoplasm

[1 mark]

ii What is the job of ribosomes?

.... **[1 mark]**

D–C

2 a The diagram shows part of a DNA molecule.

i On the diagram, put a circle around **one** base. **[1 mark]**

ii What term is usually used to describe the shape of a DNA molecule?

Double Helix

[1 mark]

iii Cells make copies of sections of DNA. These copies then pass out of the nucleus into the cytoplasm. Explain why this happens.

DNA cannot leave so the copy one leaves so proteins can be made

[2 marks]

B–A*

b The DNA bases code for amino acids.

i How many amino acids are coded for in the following section of DNA?

TATATGTAAAAACAA

5

[1 mark]

ii Write down the complementary DNA base sequence for this section of DNA.

.... **[1 mark]**

D–C

3 A scientist called Erwin Chargaff worked out the percentage of the four bases A, G, C and T in different organisms. The table shows some of his results.

	Percentage of each DNA base			
	A	G	C	T
Humans	29	21	21	29
Bacteria	24	26	26	24

a Describe the patterns shown by the scientist's results.

A is equal to T and G is equal to C.

[2 marks]

B–A*

b Explain why this data was important to Watson and Crick when they tried to work out the structure of DNA.

....

.... **[2 marks]**

Proteins and mutations

1 a Draw straight lines to join each protein to its correct function.

Protein	Function
collagen	a carrier protein
haemoglobin	a hormone
insulin	a structural protein

[2 marks]

D–C

b Write down the name of the subunits that make up protein molecules.

Amino acids [1 mark]

c Haemoglobin is called a globular protein because the molecules have a compact rounded shape. Collagen is a fibrous protein because the molecules are long and straight. Explain what causes these two proteins to have different shapes.

It has a different order of amino acids [1 mark]

B–A*

2 a The enzyme amylase breaks down starch. Explain why amylase would not break down proteins.

[2 marks]

b Look at the graph. It shows the effect of temperature on the enzyme amylase.

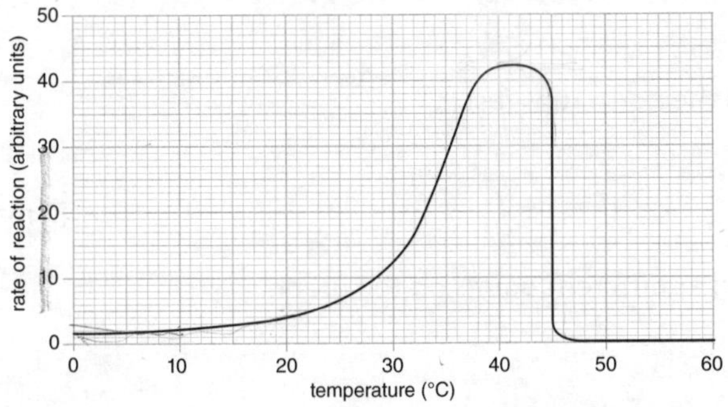

rate of reaction (arbitrary units)
temperature (°C)

i Describe the pattern shown in the graph.

[2 marks]

D–C

ii Write down the optimum temperature for this enzyme.

[1 mark]

iii Work out the Q_{10} for this enzyme between 10 and 20 °C. Answer = [2 marks]

iv Explain why the enzyme shows such an increase in rate between these temperatures.

It's a biological catalyst which speeds up the reaction

[2 marks]

B–A*

3 Changes to genes can happen spontaneously or are caused by factors in the environment.

a Write down one environmental factor that increases the chance of a change in a gene.

Mutation [1 mark]

D–C

b Why are changes to genes important for evolution?

[2 marks]

c The diagram shows how a white pigment is turned into a purple pigment in the petals of a flower using two enzymes. Explain how a change to the DNA could produce plants with red flowers.

white pigment ⇨ (Enzyme A) red pigment ⇨ (Enzyme B) purple pigment

[3 marks]

B–A*

Respiration

1 a A horse is waiting to run in a race. During the race the horse will need to generate more ATP.

Explain the function of ATP.

It's used as an energy source

[2 marks]

b The horse starts to run.

Complete the balanced symbol equation for aerobic respiration in the horse.

$C_6H_{12}O_6$ + ...$6O_2$...... → ......$6CO_2$...... + ...$6H_2O$...... **[2 marks]**

c The graph shows the lactic acid concentration in the horse's blood as it runs at different speeds.

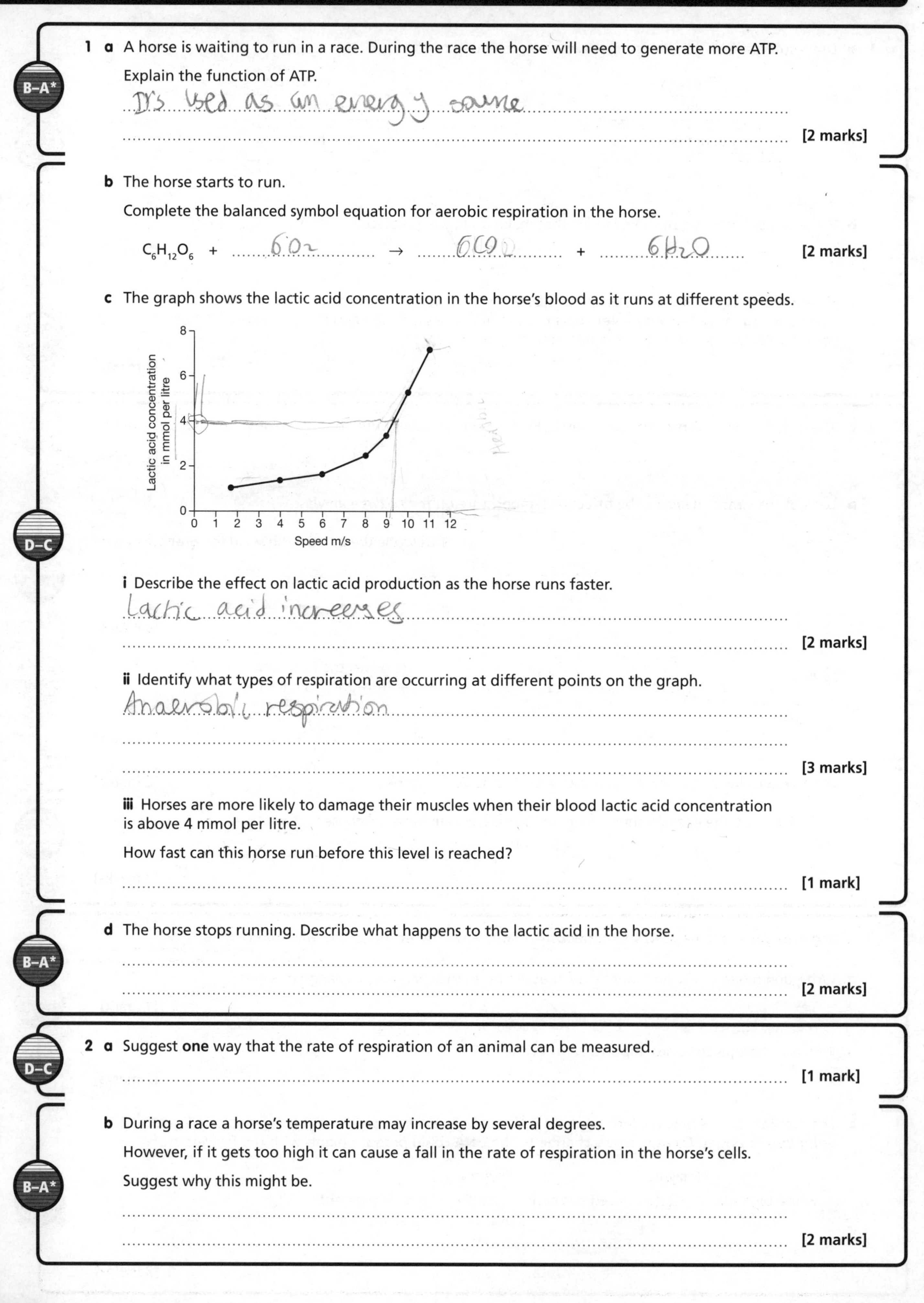

i Describe the effect on lactic acid production as the horse runs faster.

Lactic acid increases

[2 marks]

ii Identify what types of respiration are occurring at different points on the graph.

Anaerobic respiration

[3 marks]

iii Horses are more likely to damage their muscles when their blood lactic acid concentration is above 4 mmol per litre.

How fast can this horse run before this level is reached?

[1 mark]

d The horse stops running. Describe what happens to the lactic acid in the horse.

[2 marks]

2 a Suggest **one** way that the rate of respiration of an animal can be measured.

[1 mark]

b During a race a horse's temperature may increase by several degrees.

However, if it gets too high it can cause a fall in the rate of respiration in the horse's cells.

Suggest why this might be.

[2 marks]

Cell division

1 a Humans are multicellular organisms. There can be advantages to being multicellular rather than unicellular. Explain why.

....

.... **[2 marks]**

D–C

b The table shows the surface area and volume of different cubes.

i Finish the table by calculating the surface area to volume ratio of each cube. The first one has been done for you.

Cube	Surface area in cm^2	Volume in cm^3	Ratio
A	24	8	$\frac{24}{8} = 3.0$
B	54	27	
C	96	64	
D	150	125	

[3 marks]

B–A*

ii Explain why larger, multicellular organisms need to develop special exchange surfaces, such as lungs. Use data from the table to help you.

....

.... **[2 marks]**

2 Look at the statements about cell division.

a Put a tick (✓) next to each statement that refers to the type of cell division that makes new body cells.

The new cells are diploid. ☐

Four new cells are made. ☐

The new cells contain 23 chromosomes. ☐

The new cells show variation. ☐

Before cells divide, DNA replication takes place. ☐

[2 marks]

D–C

b DNA replication is called 'semiconservative'. This is because two new DNA molecules are produced and each one has one new strand and one original strand of the DNA molecule.

Explain how DNA replication produces this result.

....

.... **[2 marks]**

B–A*

3 a Scientists have discovered a mutation in the DNA of mice. They have found that this change makes the mice produce sperm without an acrosome.

The sperm that are produced without an acrosome **cannot** fertilise an egg. Explain why.

....

.... **[2 marks]**

D–C

b The diagram shows a cell dividing. The cell is shown during the first division of meiosis and during the second division.

Describe the differences between the two diagrams.

....

....

.... **[3 marks]**

B–A*

The circulatory system

D–C

1 a Red blood cells are adapted to do their job. They are disc shaped and have no nucleus. Explain how these adaptations help them do their job.

Disc shaped: .... **[1 mark]**

No nucleus: .... **[1 mark]**

b Write down the name of the chemical that makes red blood cells red.

.... **[1 mark]**

B–A*

c Explain how this chemical transports oxygen around the body.

....

.... **[3 marks]**

D–C

2 Three different types of blood vessels transport blood around the body.

a Describe the role of these blood vessels in circulating blood around the body.

....

....

.... **[3 marks]**

B–A*

b The diagrams show the three different types of blood vessels.

A	B	C

i Write the name of each type of blood vessel in the box under the correct diagram. **[2 marks]**

ii Describe how blood vessel **C** is adapted for its function.

....

.... **[2 marks]**

D–C

3 Look at the diagram of the heart.

a On the diagram of the heart, label the bicuspid valve and the aorta. **[2 marks]**

b The left ventricle has a thicker wall than the right ventricle. Explain why.

....

.... **[2 marks]**

B–A*

c The human heart is part of a double circulatory system. Write down one advantage of having a double circulatory system rather than a single system.

.... **[1 mark]**

Growth and development

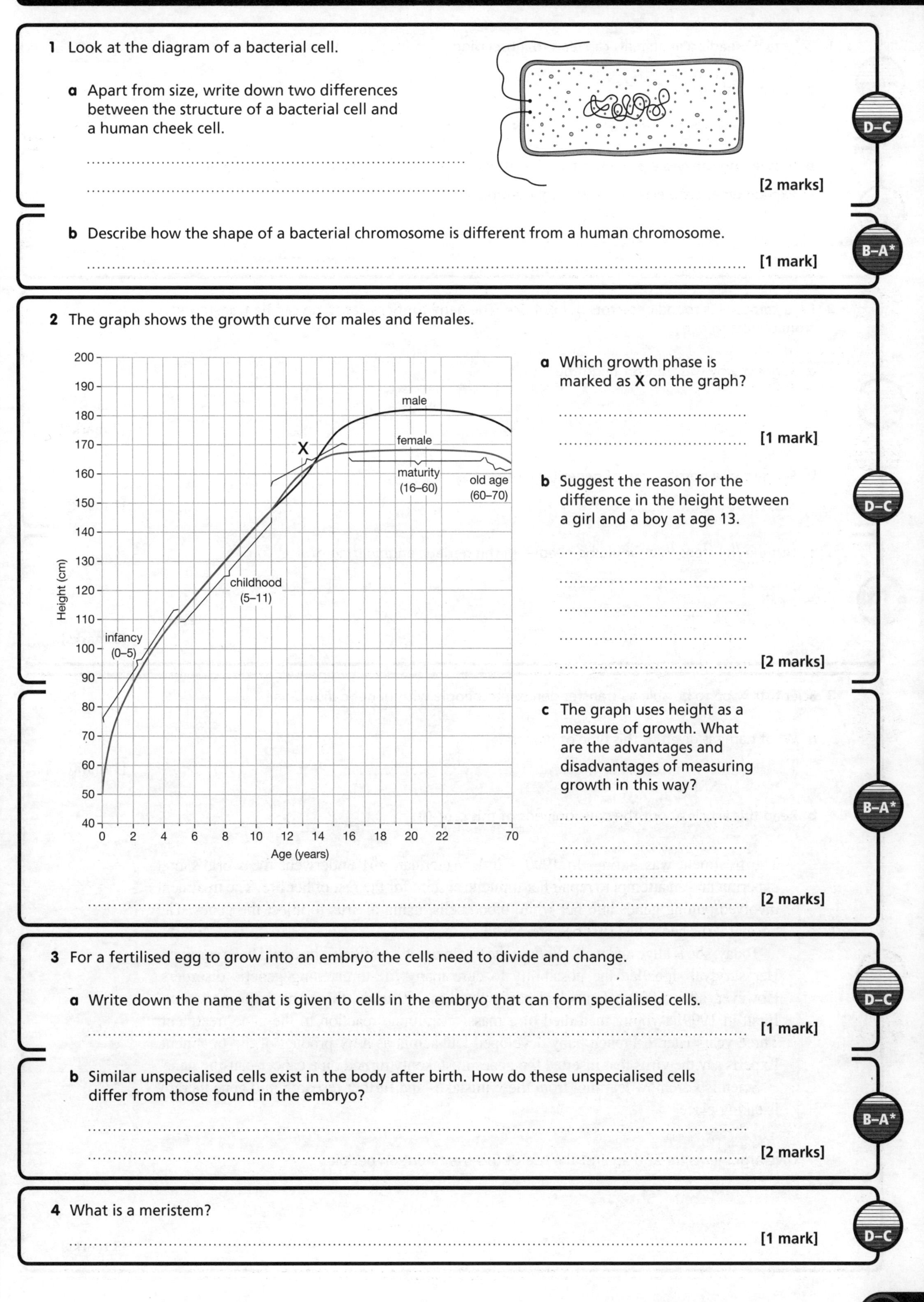

1 Look at the diagram of a bacterial cell.

a Apart from size, write down two differences between the structure of a bacterial cell and a human cheek cell.

....

....

[2 marks]

D–C

b Describe how the shape of a bacterial chromosome is different from a human chromosome.

.... **[1 mark]**

B–A*

2 The graph shows the growth curve for males and females.

a Which growth phase is marked as **X** on the graph?

....

.... **[1 mark]**

b Suggest the reason for the difference in the height between a girl and a boy at age 13.

....

....

....

....

.... **[2 marks]**

D–C

c The graph uses height as a measure of growth. What are the advantages and disadvantages of measuring growth in this way?

....

....

....

.... **[2 marks]**

B–A*

3 For a fertilised egg to grow into an embryo the cells need to divide and change.

a Write down the name that is given to cells in the embryo that can form specialised cells.

.... **[1 mark]**

D–C

b Similar unspecialised cells exist in the body after birth. How do these unspecialised cells differ from those found in the embryo?

....

.... **[2 marks]**

B–A*

4 What is a meristem?

.... **[1 mark]**

D–C

New genes for old

D–C

1 Selective breeding in animals can lead to inbreeding.

a What is inbreeding?

.... [1 mark]

B–A*

b Inbreeding can cause problems for scientists who are trying to mate endangered animals.

Explain what problems occur from inbreeding.

....

.... [2 marks]

D–C

2 Beta-carotene is found in carrots but not rice. The gene for beta-carotene can be transferred from carrots to rice.

a Suggest why this transfer might be useful.

....

.... [2 marks]

b Suggest one possible risk of genetic engineering.

.... [1 mark]

B–A*

c Outline the steps that would be needed in this genetic engineering of rice.

....

....

.... [3 marks]

D–C

3 Scientists hope to be able to transfer genes into people with genetic disorders.

a What name is given to this type of process?

.... [1 mark]

b Read this article about the development of this process.

> The treatment was daring. In 1990 a little American girl underwent the world's first experiment – an attempt to repair her immune system for the rest of her life. The treatment involved putting genes into her white blood cells, using a virus to inject the genes. The normal virus genes had first been removed.
>
> Today, she is alive and well. Without the treatment it is unlikely she could have survived. Her survival signalled the possibility to cure many life-threatening genetic disorders. However, after 5,000 patients had participated in 350 trials, things began to go wrong. First, in 1999, a young man died of a massive immune reaction to the gene treatment. Three years later, a French baby developed leukaemia as a by-product of the treatment. Experts say the virus that inserted the genes mistakenly turned on a cancer-causing gene.
>
> Scientists are now learning from these mistakes and further research into gene therapy is taking place.

Give arguments for and against the use of this treatment on people.

....

....

.... [3 marks]

Cloning

1 a Cows can be cloned using **nuclear transfer**.

What is meant by nuclear transfer?

....

.... **[2 marks]**

b Scientists are hoping to solve organ transplant problems by cloning pigs.

D–C

i Suggest how the cloning of pigs could help solve organ transplant problems.

....

.... **[2 marks]**

ii Suggest one other reason that people might want to clone pigs.

.... **[1 marks]**

2 The diagram shows a planned possible technique to produce a human cloned embryo for therapeutic use. The technique is similar to that used to produce Dolly the sheep.

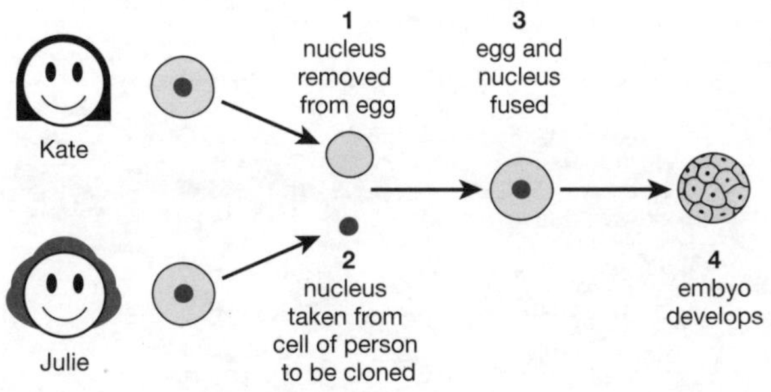

B–A*

a What is done between steps 3 and 4 to make the cell divide?

.... **[1 mark]**

b If the embryo was allowed to grow into a baby, would it most resemble Kate, or Julie? Explain your answer.

....

.... **[2 marks]**

3 Strawberry plants can reproduce sexually or asexually.

a Write down one advantage and one disadvantage to a gardener of reproducing strawberry plants asexually.

D–C

....

.... **[2 marks]**

b Plants like strawberries can be artificially reproduced by tissue culture.

Describe what is meant by the terms 'aseptic technique' and 'growth medium' in this process.

B–A*

....

.... **[2 marks]**

B3 Extended response question

Collagen is an important protein in the body.

Sometimes, disorders involving collagen cause problems in the skin and joints. They can also cause blood vessels to weaken and arteries may burst. These disorders can be inherited.

Explain how an inherited disorder can affect collagen and suggest why it causes these particular problems.

❗ The quality of written communication will be assessed in your answer to this question.

[6 marks]

Ecology in the local environment

1 Look at the kite diagram showing the distribution of different species of barnacles on rocks on a sea shore.

4 3.5 3 2.5 2 1.5 1 0.5 0

30 20 10 0 −10 −20 −30   30 20 10 0 −10 −20 −30   30 20 10 0 −10 −20 −30   30 20 10 0 −10 −20 −30

A. modestus   C. montagui   C. stellatus   S. balanoides

D–C

a Which species is most abundant?

.... **[1 mark]**

b What is this gradual change in distribution called?

....

.... **[1 mark]**

B–A*

c Suggest what abiotic factors could cause this distribution.

....

....

....

.... **[2 marks]**

D–C

2 a Compare the biodiversity of natural and artificial ecosystems. Include an example of each type.

....

....

.... **[4 marks]**

B–A*

b Natural ecosystems are self-supporting, apart from an energy source.

Explain how they are self-supporting.

....

.... **[4 marks]**

D–C

3 Charlie uses a pooter to catch ladybirds in his garden. He catches 50 and marks them with a white dot and releases them. The next day he catches 60 ladybirds, but only 10 have a white dot.

a What is the term used to describe this method of estimating population size?

.... **[1 mark]**

b i Calculate the estimated population size using the formula

$$\text{population size} = \frac{\text{number in 1st sample} \times \text{number in 2nd sample}}{\text{number in second sample previously caught}}$$

....

.... **[2 marks]**

ii Why is this only an estimate?

.... **[1 mark]**

B–A*

c What assumptions need to be made when using this method of estimating population size?

....

.... **[3 marks]**

Photosynthesis

D–C

1 a i Complete the symbol equation for photosynthesis.

$6CO_2 + 6H_2O \rightarrow$ .... **[2 marks]**

ii What is the energy source for photosynthesis?

.... **[1 mark]**

b i Complete the table to show what happens to glucose in a plant.

Changed to:	**Used for:**
cellulose	
	storage
proteins	
fats and oils	

[4 marks]

ii If glucose is not changed to other substances, what is it used for?

.... **[1 mark]**

D–C

2 a What was Priestley's important contribution in understanding photosynthesis?

.... **[1 mark]**

B–A*

b Explain how the use of isotopes changed our understanding of photosynthesis.

....

.... **[3 marks]**

D–C

3 Look at the diagram of a greenhouse.

Describe how the conditions necessary for photosynthesis are achieved.

....

....

....

....

....

....

.... **[3 marks]**

B–A*

4 The levels of oxygen and carbon dioxide in a water tank with plants and goldfish were recorded over 24 hours.

a Describe and explain the results.

....

.... **[3 marks]**

b Suggest how and why the levels would change if the plants and goldfish were removed.

....

.... **[3 marks]**

Leaves and photosynthesis

1 a Complete the table showing leaf adaptations and their uses.

Adaptation:	**Uses:**
broad leaves	increase surface area to get more light
thin leaves	diffusion
	to absorb light from different parts of the spectrum
vascular bundles	support and transport
guard cells	

[4 marks]

D–C

b An insect called a leaf miner burrows through the inside of leaves and eats the cells.

i Name two types of cells it will eat.

.... **[2 marks]**

ii Explain why the insect will cause colourless lines on the leaves.

.... **[2 marks]**

2 Look at the graph. It shows the absorption spectrum of two plant pigments.

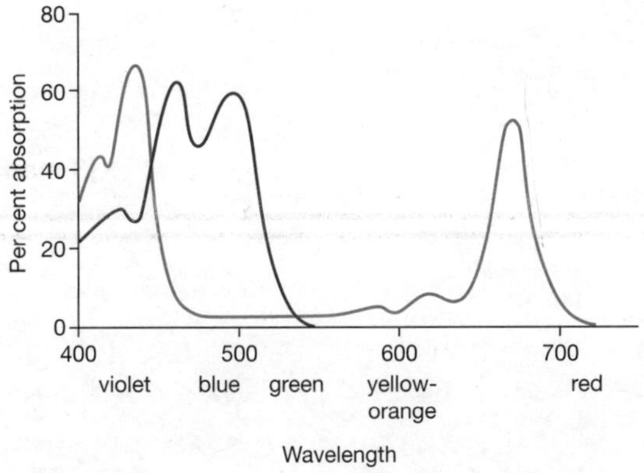

a Name two photosynthetic pigments.

....

.... **[1 mark]**

D–C

b What do you understand by the term 'absorption spectrum'?

....

....

.... **[1 mark]**

c How does the information in the graph show that the leaves maximise the use of energy from the Sun?

....

.... **[2 marks]**

B–A*

d Which part of the light spectrum is not used in photosynthesis?

.... **[1 mark]**

e If you wanted the maximum growth in your plants, which wavelengths of light would you shine on them?

.... **[2 marks]**

f The longer wavelengths of light are absorbed or reflected by the sea. Suggest what effect this will have on plants growing on the sea bed.

....

....

.... **[3 marks]**

Diffusion and osmosis

D–C

1 Ruby opens a bottle of perfume. The perfume evaporates. After a few seconds, Ruby smells the perfume.

She draws a diagram to explain this.

a Using the same symbols, draw in the second box the position of molecules where you expect them to be after a few seconds. **[2 marks]**

b Name the process involved in these changes.

.... **[1 mark]**

c Explain how differences in concentration cause these changes.

....

.... **[2 marks]**

B–A*

d Describe three ways of increasing the rate of these changes.

1 ....

2 ....

3 .... **[3 marks]**

D–C

2 Complete the definition of osmosis.

Osmosis is the movement of .... across a

.... membrane.

It takes place from an area of a .... solution to an area of a

.... solution.

B–A*

The movement is a consequence of the .... movement of individual particles. **[5 marks]**

B–A*

3 When placed in very salty water, plant cells become plasmolysed.

a Describe what happens to the contents of the plant cells.

....

.... **[2 marks]**

b What are the possible consequences of plasmolysis in plant cells?

....

.... **[2 marks]**

c Animal cells behave differently when placed in very salty water. Explain why.

....

.... **[2 marks]**

Transport in plants

1 **a** Complete the sentences about transport of water and food in plants.

Water and food travel through plants inside .... bundles.

Water travels inside .... cells.

Food travels inside .... cells. **[3 marks]**

D–C

b Describe how xylem vessels are different from phloem cells.

....

.... **[3 marks]**

B–A*

2 Look at the diagram of the lower leaf surface of two leaves kept in different conditions.

stoma closed

stoma open

leaf A

leaf B

a Count the number of open stomata in Leaf A and Leaf B.

Open stomata in Leaf A: ....

Open stomata in Leaf B: .... **[1 mark]**

b Which leaf has been kept in dark conditions? Explain your answer.

....

.... **[2 marks]**

D–C

c Explain how the plant can change its stomatal apertures.

....

....

.... **[3 marks]**

B–A*

3 Four leaves from the same plant were smeared with petroleum jelly in different ways, weighed and suspended from a string. The experiment was left for two hours and the leaves were reweighed.

Leaf	Treatment	Initial weight g	Final weight g	Weight change g
1	No petroleum jelly used	10.9	7.2	3.7
2	Petroleum jelly on both leaf surfaces	9.8	9.2	0.6
3	Petroleum jelly on lower surface	10.1	9.1	1.0
4	Petroleum jelly on upper surface	9.9	7.2	2.7

a What was the main cause of weight loss?

.... **[1 mark]**

b Use the data to show that most stomata are in the lower surface.

....

.... **[2 marks]**

D–C

c Suggest and explain the expected results if Leaf 1 was kept at a slightly higher temperature.

....

....

.... **[3 marks]**

B–A*

Plants need minerals

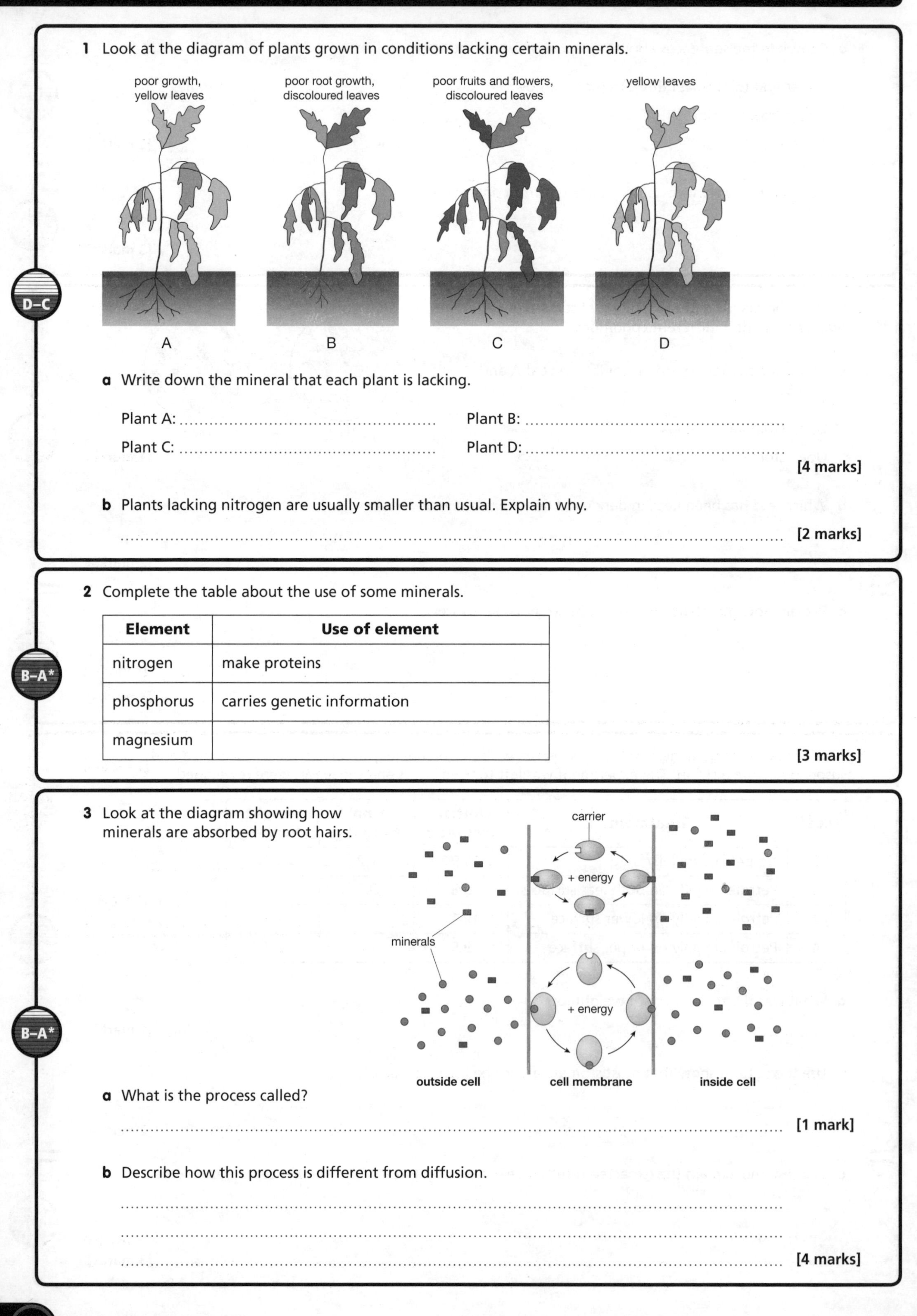

1 Look at the diagram of plants grown in conditions lacking certain minerals.

a Write down the mineral that each plant is lacking.

Plant A: .... Plant B: ....

Plant C: .... Plant D: ....

[4 marks]

b Plants lacking nitrogen are usually smaller than usual. Explain why.

.... **[2 marks]**

2 Complete the table about the use of some minerals.

Element	Use of element
nitrogen	make proteins
phosphorus	carries genetic information
magnesium	

[3 marks]

3 Look at the diagram showing how minerals are absorbed by root hairs.

a What is the process called?

.... **[1 mark]**

b Describe how this process is different from diffusion.

....

....

.... **[4 marks]**

Decay

1 Ali wants to make garden compost. He wants his garden waste to decay quickly.

a What conditions should Ali provide?

Put a tick (✓) in the box next to each correct answer.

A temperature of 25 °C ☐

A temperature of 75 °C ☐

A large amount of water ☐

Plenty of carbon dioxide ☐

Plenty of oxygen ☐ **[2 marks]**

D–C

b Name **three** useful animals he would expect to find in his compost heap.

1 ....

2 ....

3 .... **[3 marks]**

c Suggest different views people may have on having compost heaps in their garden.

....

.... **[2 marks]**

d i Explain why detritivores are important in decay.

....

.... **[2 marks]**

ii Explain how saprophytes are important in decay.

....

.... **[2 marks]**

B–A*

e Explain how and why temperature affects the rate of decay.

....

....

.... **[4 marks]**

2 Ali uses a refrigerator and freezer for his garden produce.

Explain how putting foods in:

a a refrigerator at 5 °C

b a deep freeze at –22 °C

D–C

will help to keep them for longer.

....

....

....

.... **[4 marks]**

Farming

D–C

1 Plants can be grown without the use of soil.

a What is this system of intensive farming called?

.... **[1 mark]**

B–A*

b i Explain the **advantages** of this system.

....

.... **[2 marks]**

ii Explain the **disadvantages** of this system.

....

.... **[2 marks]**

D–C

2 In 1935 large cane toads from South America were introduced into Queensland in Australia to control insect pests that attack sugar cane plants.

The toads had little effect on the insect pest population. They have now spread into nearly all areas of Australia and have no natural predators. They are a pest, eating the smaller native toad and causing a great deal of damage. Because of their large size and poisonous skin, they have few predators.

The insect pests are now controlled by insecticides.

a This is an example of introducing an organism into an ecosystem to control another organism. What is this method called?

.... **[1 mark]**

b Suggest why the introduction of cane toads was thought to be better than using insecticides.

....

.... **[2 marks]**

c Suggest why the introduction of cane toads was not a success.

....

.... **[2 marks]**

D–C

3 The use of crop rotation is important in farming.

a Some farmers use a crop rotation system. Explain why.

....

.... **[2 marks]**

b Some organic farmers vary the seed planting time of their crops. Explain why.

....

.... **[2 marks]**

B4 Extended response question

Materials can enter and leave cells.

The diagram represents a plant cell in water.

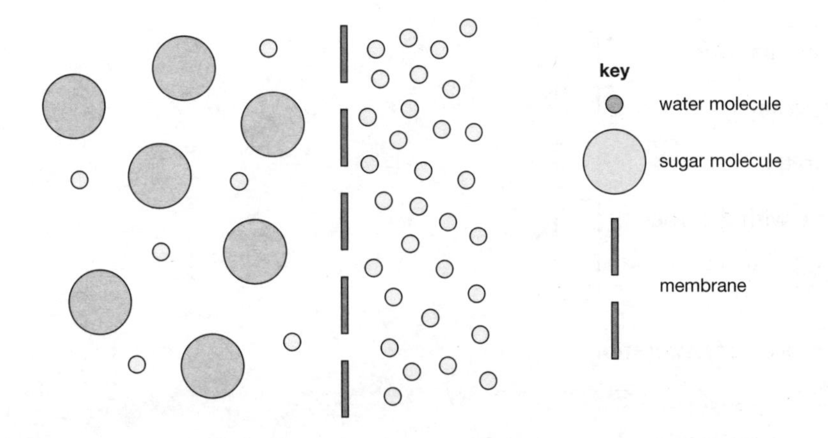

Explain your predicted movement of molecules and the importance of any movement.

The quality of written communication will be assessed in your answer to this question.

....

[6 marks]

Skeletons

D–C

1 a Look at the statements about long bones. Put a tick (✓) in the boxes next to the three correct statements.

Long bones:

- contain bone marrow ✓
- are hollow in adults ☐
- are found in the rib cage ☐
- cannot grow with the body ☐
- contain living cells ☐

[3 marks]

B–A*

b Describe the process of ossification.

Ossification is when the long bone grows with a bone marrow in it and the cartilage on the head of it.

[4 marks]

D–C

2 Look at the diagram of the bones and muscles in a forearm.

a What type of synovial joint is found at the:

i shoulder?

Ball and socket

ii elbow?

Hinge

[2 marks]

B–A*

b What are the functions of the synovial fluid?

Acts as a cushion and allows easy movement

[2 marks]

D–C

c Describe the actions of antagonistic muscles in the movement of the forearm.

Bicep muscle contracts and tricep muscle relaxes as arm is pushed up. Bicep muscle relaxes and tricep muscle contracts as arm relaxes

[4 marks]

B–A*

d Explain how the forearm acts as a lever.

The arm is raised up and lowers down

[2 marks]

B–A*

3 Describe the causes and consequences of osteoporosis.

Lack of calcium and phosphates can lead to fracture.

[3 marks]

Circulatory systems and the cardiac cycle

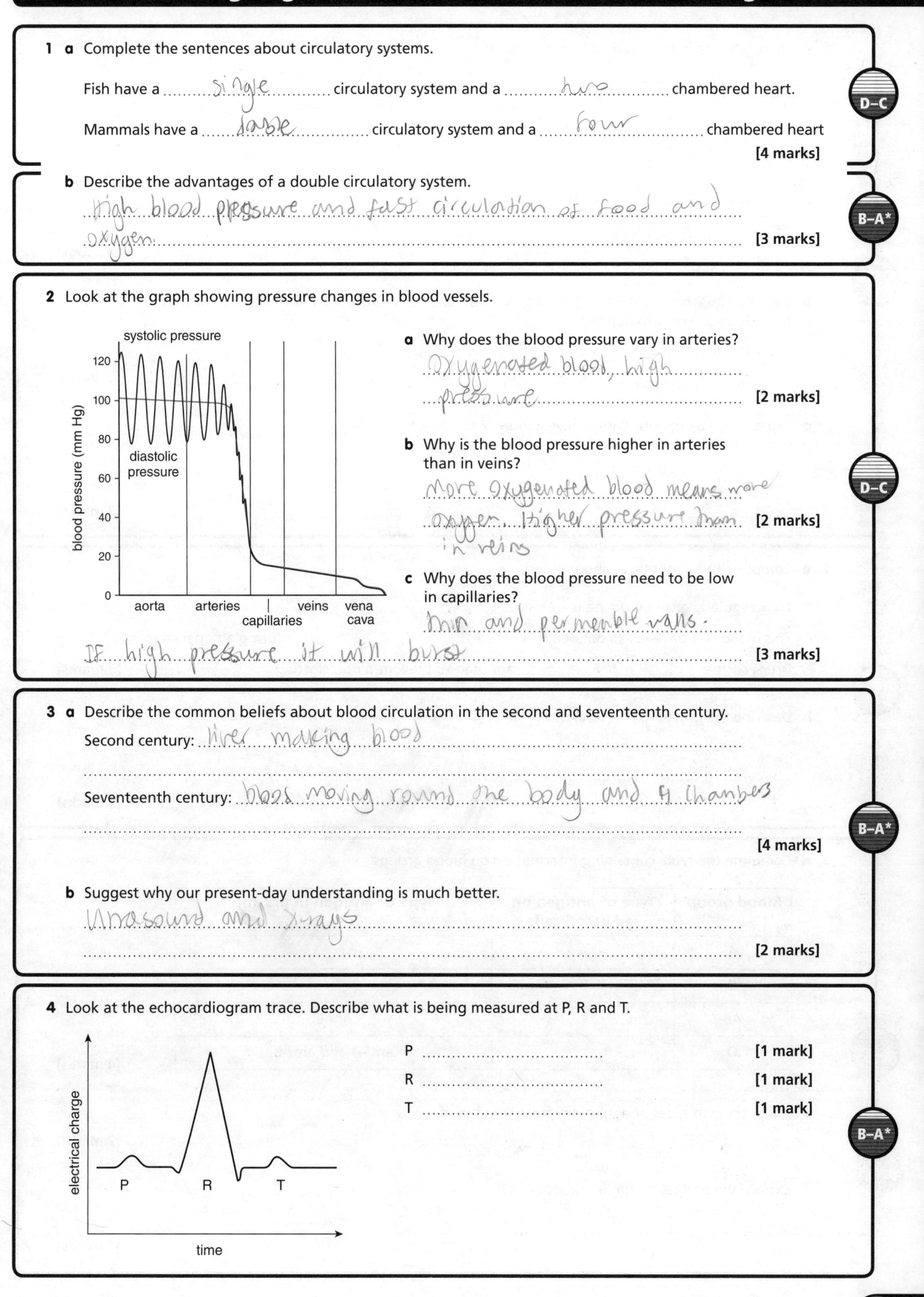

1 a Complete the sentences about circulatory systems.

Fish have a ....single........ circulatory system and a ....two........ chambered heart.

Mammals have a ....double........ circulatory system and a ....four........ chambered heart

[4 marks]

D–C

b Describe the advantages of a double circulatory system.

High blood pressure and fast circulation of food and oxygen.

[3 marks]

B–A*

2 Look at the graph showing pressure changes in blood vessels.

a Why does the blood pressure vary in arteries?

Oxygenated blood, high pressure

[2 marks]

b Why is the blood pressure higher in arteries than in veins?

More oxygenated blood means more oxygen. Higher pressure than in veins

[2 marks]

D–C

c Why does the blood pressure need to be low in capillaries?

thin and permeable walls.

If high pressure it will burst

[3 marks]

3 a Describe the common beliefs about blood circulation in the second and seventeenth century.

Second century: liver making blood

Seventeenth century: blood moving round the body and 4 chambers

[4 marks]

B–A*

b Suggest why our present-day understanding is much better.

Ultrasound and X-rays

[2 marks]

4 Look at the echocardiogram trace. Describe what is being measured at P, R and T.

P .... **[1 mark]**

R .... **[1 mark]**

T .... **[1 mark]**

B–A*

Running repairs

1 Some babies are born with a 'hole in the heart'.

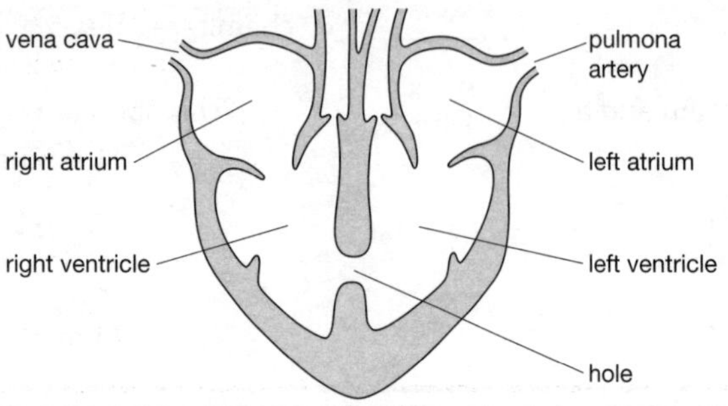

a Explain the consequences to the blood circulation of this condition.

....

....

....

....

.... **[3 marks]**

B–A*

b Babies with a 'hole in the heart' are called 'blue babies' because their skin is bluish rather than pink. Suggest why this happens.

....

.... **[2 marks]**

c Explain why all unborn babies have a 'hole in the heart'.

....

....

.... **[3 marks]**

2 **a** Complete these sentences about blood.

Tony regularly gives blood, he is a blood ....

This blood will be ready to be used in a blood .... during an operation.

Drugs such as .... are used to prevent blood clotting. **[3 marks]**

D–C

b Describe the process of blood clotting.

....

....

.... **[3 marks]**

3 **a** Complete the table containing information on blood groups.

Blood group	Type of antigen on red blood cells	Type of antigen in plasma
A	A	
B	B	
AB		none
O		anti-A and anti-B

[4 marks]

B–A*

b Name the two types of agglutinin present in blood.

.... **[2 marks]**

c Explain the process of agglutination.

....

.... **[3 marks]**

Respiratory systems

1 a Draw straight lines to join up each condition/disease with its correct cause and description.

Cause	Condition/disease	Description
lifestyle choice	asbestosis	too much mucus in bronchioles
genetic	cystic fibrosis	inflammation/scarring of lungs
exposure to fibres in factory	lung cancer	rapid cell growth

[4 marks]

D–C

b Describe how the respiratory system protects itself against disease.

....

.... **[2 marks]**

c Describe the symptoms of asthma.

....

.... **[3 marks]**

d Describe what happens during an asthma attack.

....

.... **[3 marks]**

B–A*

2 Look at the spirometer trace.

Using this trace, describe and explain the effects of exercise on breathing.

....

....

....

....

.... **[3 marks]**

D–C

3 The apparatus shown in the diagram can be used to demonstrate the action of the diaphragm during breathing.

a In Diagram B, what will happen to the air volume and pressure inside the bell jar?

....

....

....

.... **[2 marks]**

D–C

b In Diagram B, draw in the expected shape of the balloon. **[1 mark]**

c Suggest one reason why the model does not fully explain how the air is breathed in and out.

.... **[1 mark]**

Digestion

D–C

1 Food is digested physically and chemically. Explain the importance of physical digestion.

....

....

.... **[3 marks]**

D–C

2 In 1822, a young man called Alexis was shot in the stomach. The wound did not heal but he survived. His doctor did an experiment in which he placed meat in Alexis' stomach.

meat when first put into Alexis' stomach

meat after two hours in Alexis' stomach

a Look at the diagram. Describe and explain what happened to the meat.

....

....

.... **[4 marks]**

B–A*

b The pH of the stomach is different from that in the mouth and small intestine. Explain how and why it is different.

....

....

.... **[3 marks]**

D–C

3 Digested food molecules are absorbed through the walls of the small intestine.

a Which food molecules are absorbed into the blood plasma?

.... **[2 marks]**

b Which food molecules are absorbed into the lymph?

.... **[1 mark]**

c Suggest why different food molecules are absorbed into different structures.

....

.... **[3 marks]**

B–A*

4 What effect will the removal of the gall bladder have on digestion?

....

....

.... **[3 marks]**

B–A*

5 The chemical breakdown of starch is described as a two-step process. Describe the process.

....

....

.... **[4 marks]**

Waste disposal

1 a Look at the diagram of the kidney.

Name the structures A, B, C, D and E.

A ....

B ....

C ....

D ....

E ....

[5 marks]

D–C

b Describe the functions of the **three** main parts of a kidney tubule.

....

....

.... **[3 marks]**

B–A*

2 The diagram shows the contents of blood plasma and urine.

Contents of blood plasma in g/100 cm³	
water	910
protein	74
glucose	1
urea	0.3
salt	9

Contents of urine in g/100 cm³	
water	960
protein	0
glucose	0
urea	20
salt	12

a Write down **two** ways in which the contents differ.

1 ....

2 .... **[2 marks]**

b Explain these differences.

....

....

.... **[4 marks]**

D–C

c Suggest and explain what would happen to the contents of urine on a hot day.

....

.... **[2 marks]**

d Suggest **two other** factors that could affect urine concentration.

.... **[2 marks]**

3 a Carbon dioxide must be removed from the body. Explain why.

.... **[1 mark]**

D–C

b Explain how the body responds to increased carbon dioxide levels.

....

.... **[2 marks]**

B–A*

4 Explain the role of a negative feedback mechanism in regulating the concentration of urine.

....

....

....

.... **[4 marks]**

B–A*

Life goes on

1 a Draw lines to link up each hormone to its correct function.

Hormone	Function
oestrogen	controls ovulation
progesterone	stimulates an egg to develop
LH	causes repair of uterus wall
FSH	maintains uterus wall

[3 marks]

D–C

b Look at the diagram that shows levels of oestrogen and progesterone in two menstrual cycles. Describe and explain the difference in the levels of oestrogen and progesterone before and after fertilisation.

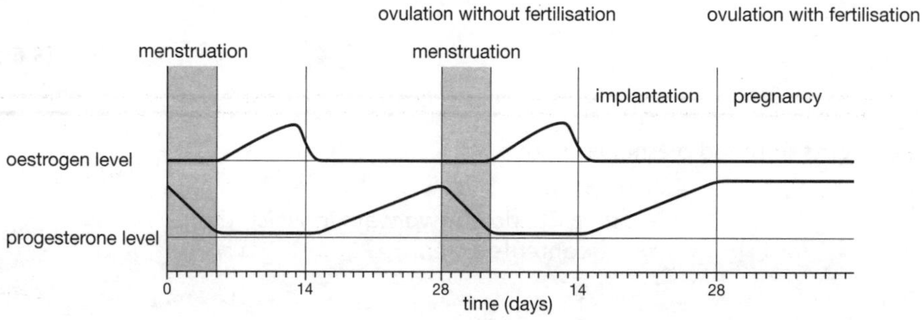

ovulation without fertilisation
ovulation with fertilisation
menstruation
menstruation
implantation
pregnancy
oestrogen level
progesterone level
0
14
28
14
28
time (days)

....

....

.... [4 marks]

B–A*

c Explain how a negative feedback system controls oestrogen levels.

....

....

.... [3 marks]

2 Complete the table of information about infertility treatments provided by one clinic.

B–A*

Treatment	Success rate	Risks/disadvantages
IVF	34%	straightforward technique but risk of ....
egg donation	50%	carries father's genetic code but ....
artificial insemination	24%	lower success rates than IVF because ....
surrogacy	too few to be reliable	surrogate mother might ....
use of FSH	not used independently	difficult to judge amount and timings of ....
ovary transplant	too few to be reliable	problems with ....

[5 marks]

Growth and repair

1 The world's first arm transplant was done in 1998 to replace an arm damaged in an accident.

a What are the main problems in supply of donor organs?

....

.... **[3 marks]**

D–C

b What ethical issues were raised in this arm transplant?

....

.... **[2 marks]**

c The transplant required joining nerves, blood vessels and muscles. What are the main problems to the recipient after receiving a transplant?

....

.... **[2 marks]**

B–A*

d Many countries around the world use a donor register. Write down **two** advantages of having such a donor register.

....

.... **[? marks]**

2 Look at the data on organ donation.

Cornea	Heart	Heart and lung	Liver	Pancreas	Kidney
4115 donated	95 on transplant list	287 on transplant list	268 on transplant list	216 on transplant list	6890 on transplant list
2490 grafted	122 transplants	122 transplants	638 transplants	209 transplants	1453 transplants
	135 donors	135 donors	632 donors	135 donors	790 donors

a Which of these organ transplants can be done from a living donor?

.... **[1 mark]**

B–A*

b 'Corneal grafts (transplants) are simple operations with a high success rate.' Explain whether this statement is supported by the data.

.... **[1 mark]**

c Suggest an explanation for the link between the number of kidney donors and transplants.

....

.... **[2 marks]**

d The numbers of heart and heart and lung transplants are low. Suggest why.

.... **[1 mark]**

3 The life expectancy of people in the UK has increased in recent times. Write down **three** possible reasons for this.

....

....

.... **[3 marks]**

D–C

B5 Extended response question

A fish and a human have different blood circulatory systems.

Draw diagrams to illustrate how they are different and explain the consequences on heart structure and function.

The quality of written communication will be assessed in your answer to this question.

....

....

....

....

....

....

....

....

....

....

....

....

....

....

....

....

....

....

....

.... **[6 marks]**

Understanding microbes

1 a Bacteria can be classified by their shape. Round bacteria are called spherical. Write down the names of **two other** bacteria shapes.

.... **[2 marks]**

D–C

b Bacteria reproduce asexually by splitting in half. Write down the name of this type of special asexual reproduction.

.... **[1 mark]**

2 a Bacteria that get inside our bodies reproduce very quickly. Suggest why this is a problem.

.... **[1 mark]**

b Bacteria can be grown on an agar dish. Disposable gloves are worn while handling bacteria. Explain why.

B–A*

.... **[1 mark]**

c It is important that the lids are not left off the dishes. Explain why.

.... **[1 mark]**

3 The graph shows the growth rate of yeast at different temperatures.

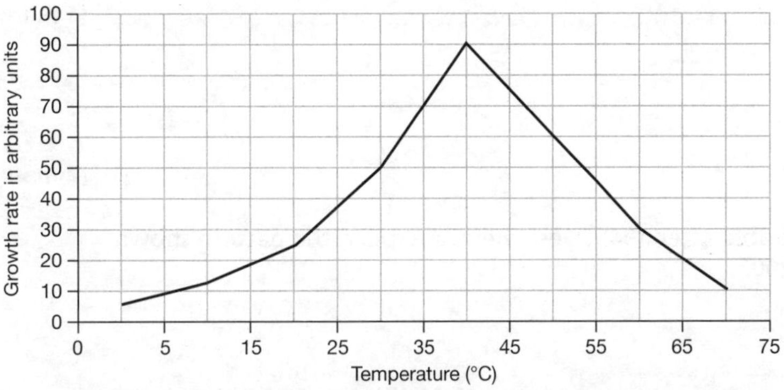

a One condition needed for optimum growth of yeast is the correct temperature. What temperature gives the optimum growth for this type of yeast?

D–C

.... **[1 mark]**

b Write down **two other** conditions needed for optimum growth.

.... **[2 marks]**

c Use information from the graph to describe how an increase from 25 °C to 35 °C will alter the growth rate of the yeast.

B–A*

....

.... **[2 marks]**

4 The diagram shows a virus particle.

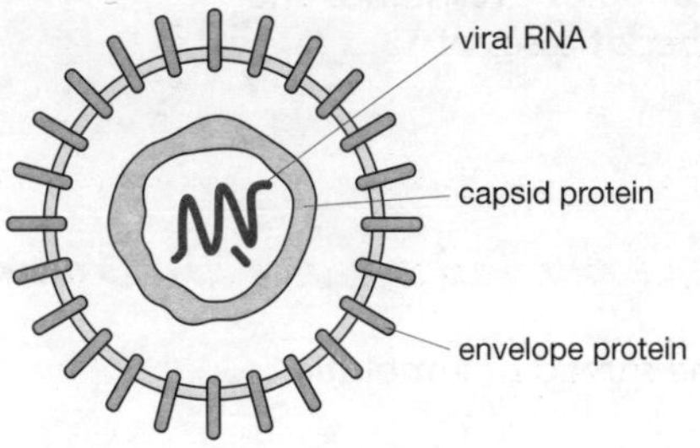

a Describe how the structure of a virus is different from that of a typical bacterium.

....

.... **[2 marks]**

B–A*

b Describe how a virus reproduces.

....

....

.... **[3 marks]**

Harmful microorganisms

D–C

1 **a** The microbe that causes the disease tuberculosis (TB) can be passed on by droplet spread. Suggest one way that this spread can be reduced.

.... **[1 mark]**

b After infection by the microorganism it takes time for the symptoms to appear. During this time the microorganisms are reproducing. Write down the term is used to describe the time between infection and the appearance of symptoms.

.... **[1 mark]**

c Look at the graph. It shows the number of TB cases reported to doctors and deaths from TB in Canada between 1924 and 2002.

One conclusion from the graph is that everybody who got TB in 1930 died from it.

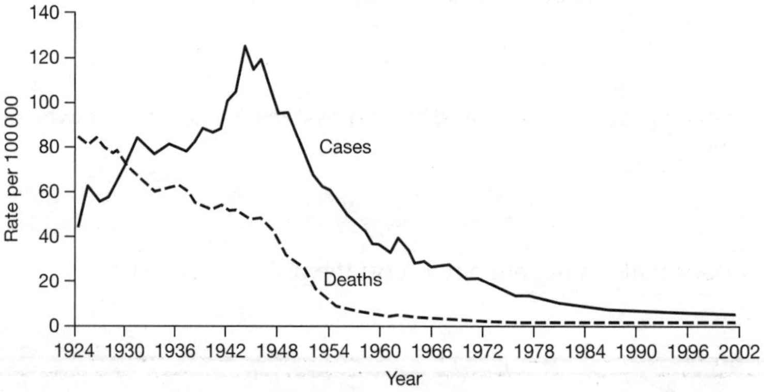

B–A*

i Why is it easy to make this conclusion from this graph?

.... **[1 mark]**

ii Why is it unlikely to be true?

....

.... **[2 marks]**

iii Antibiotics to treat TB became available after 1945. Describe and explain the pattern shown by the deaths from TB since 1924 to 2002.

....

....

.... **[3 marks]**

2 Penicillin is one antibiotic that can be used to cure bacterial infection.

D–C

a Write down the name of the scientist who discovered penicillin.

.... **[1 mark]**

b Penicillin is produced by a type of microorganism. Which type?

.... **[1 mark]**

c **i** Look back at the graph in question 1c. Explain what is meant by antibiotic resistance and describe how it explains the shape of the graph showing the number of cases of TB.

....

....

.... **[3 marks]]**

B–A*

ii Write down one precaution that can be taken to try and limit the spread of antibiotic resistant strains of TB.

.... **[1 mark]**

Useful microorganisms

1 a Karen makes some yoghurt. Here are four sentences **(A–D)** about how she makes it.

A Bacteria is added to the milk and left at about 46 °C for 4 hours.

B Milk is heated to between 78 and 90 °C then cooled.

C The yoghurt is cooled and packed.

D Flavours are added.

The sentences are in the wrong order. Fill in the boxes to show the correct order. The first one has been done for you.

B			

[1 mark]

D–C

b Describe the action of bacteria that Karen adds to the milk.

....

.... **[2 marks]**

B–A*

2 a During the process of making some beers, the beer is clarified and pasteurised. What is meant by the term clarified?

.... **[1 mark]**

b Drinks such as beer that have been made by fermentation can be modified by distillation. Describe the purpose of this distillation.

....

.... **[2 marks]**

D–C

c Why is it important to pasteurise the beer?

.... **[1 mark]**

B–A*

3 Look at the graph. It shows the growth of two different strains of yeast over time.

A

B

a The two strains of yeast are respiring glucose anaerobically to produce alcohol.

Complete the symbol equation for this process. Make sure that the equation is balanced.

.... $\rightarrow$ $2C_2H_5OH$ + ...

[2 marks]

B–A*

b The two different strains of yeast are being grown separately but under the same conditions. Describe and explain the differences in their patterns of growth.

....

....

....

.... **[4 marks]**

Biofuels

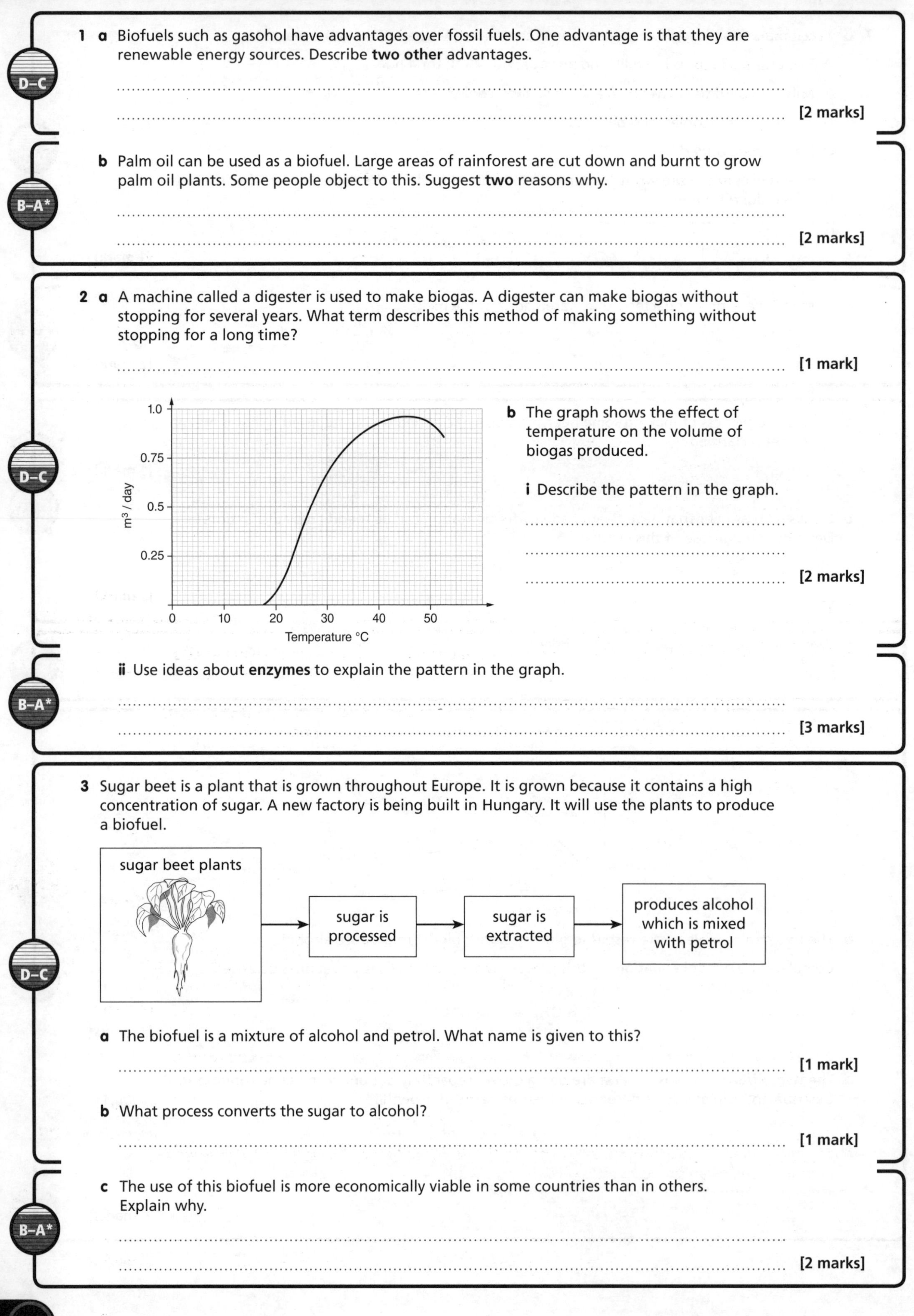

D–C

1 a Biofuels such as gasohol have advantages over fossil fuels. One advantage is that they are renewable energy sources. Describe **two other** advantages.

....

.... **[2 marks]**

B–A*

b Palm oil can be used as a biofuel. Large areas of rainforest are cut down and burnt to grow palm oil plants. Some people object to this. Suggest **two** reasons why.

....

.... **[2 marks]**

D–C

2 a A machine called a digester is used to make biogas. A digester can make biogas without stopping for several years. What term describes this method of making something without stopping for a long time?

.... **[1 mark]**

b The graph shows the effect of temperature on the volume of biogas produced.

i Describe the pattern in the graph.

....

....

.... **[2 marks]**

B–A*

ii Use ideas about **enzymes** to explain the pattern in the graph.

....

.... **[3 marks]**

D–C

3 Sugar beet is a plant that is grown throughout Europe. It is grown because it contains a high concentration of sugar. A new factory is being built in Hungary. It will use the plants to produce a biofuel.

a The biofuel is a mixture of alcohol and petrol. What name is given to this?

.... **[1 mark]**

b What process converts the sugar to alcohol?

.... **[1 mark]**

B–A*

c The use of this biofuel is more economically viable in some countries than in others. Explain why.

....

.... **[2 marks]**

Life in soil

1 Soil contains mineral particles of different sizes.

a Soil also contains dead organic material. Write down the name given to the decomposed organic material found in soil.

.... **[1 mark]**

D–C

b Describe an experiment to compare the organic material content of two different soil samples.

....

....

.... **[3 marks]**

c A soil sample is shaken with some water in a measuring cylinder. It is then left for the mineral particles to settle.

A B C D

water
silt
clay
sand
small stones

Which measuring cylinder A, B, C or D shows the correct order of the layers?

.... **[1 mark]**

B–A*

2 A group of scientists are investigating different soil treatments. They want to know what effects they have on crop yield. They treated soils in a number of ways and then measured the yield.

Treatment	Digging the soil	Digging the soil and adding compost	Digging the soil and adding worms
yield (in kg/field/year)	3000	5200	5500

a Work out the percentage increase in yield obtained by adding compost.

....

.... **[2 marks]**

D–C

b Compost contains dead organic material. How can adding this increase the yield of the soil?

....

.... **[2 marks]**

c Adding worms also increases the amount of dead organic material in the soil. How do worms achieve this?

.... **[1 mark]**

d Digging and adding earthworms also improves the drainage of the soil. Explain why a soil with poor drainage often produces poor yields.

....

....

.... **[3 marks]**

B–A*

Microscopic life in water

D–C

1 a Adult frogs develop from young called tadpoles. Tadpoles live in water.

There are advantages and disadvantages to living in water. Write down **one** advantage and **one** disadvantage.

....

.... **[2 marks]**

B–A*

b Amoeba are microscopic animals that live in fresh water.

i The water they live in is more dilute than their cytoplasm. Explain why this is a problem.

....

.... **[2 marks]**

ii Describe the process amoebas use to solve the problem. You may use labelled diagrams (on separate paper) in your answer.

....

.... **[2 marks]**

D–C

2 Read the following article about life deep in the oceans.

Organisms living deep in the ocean must deal with extreme conditions. Very few organisms were thought to be able to live there. These ideas changed dramatically in the 1970s when the first hydrothermal vents were discovered in the Pacific Ocean. These are areas where water is heated by the Earth's crust and is ejected into the ocean.

What was truly amazing about hydrothermal vents was that clustered around them were dense colonies of large animals such as tube worms, and crabs that fed on the worms. Plants cannot grow as deep as the vents and little marine snow reaches them, so scientists were puzzled about what the worms were eating. The sea water near vents is full of hydrogen sulfide, carbon dioxide and oxygen. It was discovered that special bacteria could live on the hydrogen sulfide, and that these bacteria form the base of the food chain.

a Explain why plants cannot be the source of food at the hydrothermal vents.

.... **[1 mark]**

B–A*

b Write a food chain showing the feeding relationships at the hydrothermal vents.

.... **[1 mark]**

c What is meant by marine snow?

.... **[1 mark]**

D–C

3 a Fertiliser run-off from farms can cause **eutrophication.** Describe the main stages in eutrophication. Start with fertiliser entering the water.

....

....

.... **[3 marks]**

B–A*

b Pesticides from farms can also enter the water causing problems to wildlife.

Explain why pesticides are a problem. Use ideas about food chains in your answer.

....

....

.... **[3 marks]**

Enzymes in action

1 a Different enzymes used in washing powders digest different types of stains.
The list shows some different enzymes.

Draw a straight line from each enzyme to one type of stain it digests.

amylase	starch
lipase	protein
protease	fat

[2 marks]

D–C

b Explain why digesting these stains helps to remove them from clothes.

....

.... **[2 marks]**

B–A*

2 Sucrase is an enzyme used to breakdown sucrose.

a Explain why this enzyme is useful to the food industry.

.... **[1 mark]**

D–C

b Finish the word equation to show the action of sucrase.

sucrase
sucrose → .... + ... **[2 marks]**

B–A*

3 Enzymes can be immobilised by covering them in alginate.

a Describe **two** advantages of immobilising enzymes.

....

.... **[2 marks]**

D–C

b Immobilised enzymes can be used to make lactose-free milk.

i Write down the name of the enzyme used to make lactose-free milk.

.... **[1 mark]**

ii Describe what happens in the digestive systems of people with lactose intolerance if they drink milk containing lactose.

....

....

.... **[3 marks]**

B–A*

Gene technology

D–C

1 a Scientists can change the DNA of an organism.

Write down the term that describes an organism that has had its DNA changed.

.... **[1 mark]**

b The process of changing the DNA of an organism involves several stages.

First, a section of DNA (a gene) is removed from one organism.

i Describe the next **two** stages.

....

.... **[2 marks]**

B–A*

ii Which type of enzyme is needed to remove the DNA from the organism?

.... **[1 mark]**

iii This enzyme produces 'sticky ends'.

What is a sticky end and why is it useful?

....

....

.... **[3 marks]**

D–C

2 a Human insulin can now be made using bacteria. Describe how.

....

....

....

.... **[4 marks]**

B–A*

b Describe how scientists check which bacteria have taken up the human insulin gene.

....

.... **[2 marks]**

D–C

3 a Look at the picture. It shows DNA fingerprints of individuals connected to a robbery.

Who left blood at the crime scene? Explain your answer.

....

.... **[2 marks]**

B–A*

b Describe how a DNA fingerprint is made.

....

....

.... **[3 marks]**

B6 Extended response question

Samuel wants to see if he could make wine with more alcohol by adding twice the amount of sugar than his recipe suggested.

He sets up two fermentation jars with the same amount of yeast and fruit juice but one with extra sugar added.

He then measures the concentration of alcohol in the wine every day.

His results are shown in the graph.

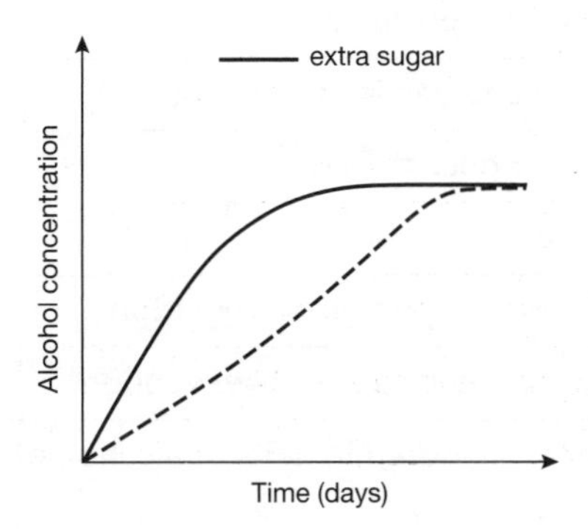

Use your knowledge of fermentation to explain the shape of the graph that Samuel obtains.

❗ The quality of written communication will be assessed in your answer to this question.

[6 marks]

B1 Grade booster checklist

B1	
I can describe the factors that increase or decrease blood pressure.	
I can explain how a diet containing high levels of saturated fats and salt can increase the risk of heart disease.	
I can explain why protein deficiency is common in developing countries.	
I can calculate a person's BMI and their EAR for protein.	
I can explain the difference between active and passive immunity.	
I can describe changes in lifestyle that may reduce the risk of some cancers.	
I can describe the path taken by a spinal reflex.	
I can explain the differences between monocular and binocular vision.	
I can explain why damage to ciliated cells can lead to a 'smoker's cough'.	
I can interpret data on the alcohol content, measured in units of alcohol, of different alcoholic drinks.	
I can describe the dangerous effects of high and low temperatures on the body.	
I can explain how Type 2 diabetes can be controlled by diet but Type 1 diabetes needs insulin doses.	
I can describe shoots as being positively phototropic and negatively geotropic.	
I can describe the commercial uses of plant hormones.	
I can describe how sex chromosomes, XX in female and XY in male, determine the sex of an individual.	
I can explain the causes of genetic variation.	
I am working at grades D–C	

I can explain the possible consequences of having high or low blood pressure.	
I can explain how narrowed coronary arteries, together with a thrombosis, increase the risk of a heart attack.	
I can describe the storage of carbohydrates (as glycogen) in the liver and fats (as adipose tissue) in the skin.	
I can describe the differences between first and second class proteins.	
I can interpret data on types of cancer and their survival/ mortality rates.	
I can describe the benefits and risks of being immunised.	
I can explain how the eye accommodates (focuses light) for distant and close objects.	
I can explain how a nerve impulse is transmitted across a synapse.	
I can explain the action of depressant and stimulant drugs on synapses.	
I can interpret information on reaction times, accident statistics and alcohol levels.	
I can explain how negative feedback mechanisms are used in homeostasis.	
I can explain how insulin helps to regulate blood sugar levels by converting excess glucose into glycogen.	
I can use information on auxins to interpret data from phototropism experiments.	
I can explain how different levels of auxin cause shoot curvature towards light.	
I can use and explain a monohybrid cross involving dominant and recessive alleles.	
I can use genetic diagrams to predict the possibilities of inherited disorders in the next generation.	
I am working at grades B–A*	

B2	
I can name organisms using the binomial system.	
I can recall the definition of a species.	
I can describe how energy is lost from food chains.	
I can explain why pyramids of energy and pyramids of biomass can be different shapes.	
I can describe how plants obtain nitrogen from the soil.	
I can describe the role of decomposers in the carbon and nitrogen cycles.	
I can explain the difference between parasitism and mutualism.	
I can explain the reasons for changes in predator and prey numbers.	
I can explain some of the adaptations that animals have for living in cold conditions.	
I can explain some of the adaptations that organisms have for living in dry conditions.	
I can use Darwin's theory of natural selection to explain how evolution occurs.	
I can state why Darwin's theory was not well accepted by many people.	
I can describe how indicator species can show levels of pollution.	
I can explain the causes and consequences of global warming.	
I can explain why it is considered important to set up conservation programmes.	
I can explain what sustainable development means.	
I am working at grades D–C	

I can describe the difference between an artificial and a natural classification system.	
I can explain that organisms can be similar because they may be closely related or because they may have evolved to live in similar environments.	
I can describe the problems involved in constructing pyramids of biomass.	
I can explain why food chains always have a limited number of trophic levels.	
I can recall the roles of all four types of bacteria in the nitrogen cycle.	
I can describe how oceans can act as carbon sinks.	
I can explain why the peak in numbers of predators and prey do not coincide.	
I can explain what is meant by an ecological niche.	
I can explain why animals in colder climates tend to be larger.	
I can describe the differences between specialists and generalists.	
I can describe the importance of reproductive isolation in speciation.	
I can explain the difference between Lamarck's theory and Darwin's theory.	
I can recall the advantages and disadvantages of using indicator species to judge pollution levels.	
I can explain why people in different countries have different carbon footprints.	
I can explain the importance of genetic variation in species.	
I can recognise some of the difficulties in trying to protect whales.	
I am working at grades B–A*	

B3 Grade booster checklist

B3	
I can explain why liver and muscle cells have large numbers of mitochondria.	
I can describe the shape of a DNA molecule.	
I can explain why enzymes are specific.	
I can describe the effect of temperature on enzyme action.	
I can recall the symbol equation for aerobic respiration.	
I can explain why anaerobic respiration occurs during exercise.	
I can explain some of the advantages of being multicellular.	
I can recall the uses of mitosis and meiosis in mammals.	
I can explain how a red blood cell is adapted to its function.	
I can label the main parts of the mammalian heart.	
I can identify simple differences between bacterial cells and plant and animal cells.	
I can recall the function of stem cells.	
I can recognise that selective breeding may lead to inbreeding.	
I can explain some potential benefits and risks of genetic engineering.	
I can describe some possible uses of cloning.	
I can describe what is meant by nuclear transfer.	
I am working at grades D–C	

I can describe the position and function of ribosomes in a cell.	
I can explain how the bases in DNA code for proteins.	
I can explain why changing temperatures affect enzyme action.	
I can explain how changes to genes can alter proteins.	
I can recall the function of ATP.	
I can explain what is meant by oxygen debt.	
I can describe the main stages in mitosis.	
I can explain how meiosis produces haploid cells.	
I can explain how haemoglobin transports oxygen.	
I can explain how the main types of blood vessels are adapted for their functions.	
I can explain the advantages and disadvantages of different measures of growth.	
I can explain the difference between adult and embryonic stem cells.	
I can explain some of the consequences of inbreeding.	
I can describe the main stages in genetic engineering.	
I can describe the cloning technique used to produce Dolly the sheep.	
I can explain why cloning plants is easier than cloning animals.	
I am working at grades B–A*	

B4 Grade booster checklist

B4	
I can calculate an estimate of a population size.	
I can compare the biodiversity of natural and artificial ecosystems.	
I can recall and use the balanced symbol equation for photosynthesis.	
I can explain why plants carry out respiration at all times.	
I can name and locate the main parts of a leaf.	
I can explain how leaves are adapted for photosynthesis.	
I can explain the net movement of particles in diffusion.	
I can describe the process of osmosis.	
I can recall that transpiration is the evaporation and diffusion of water from inside leaves	
I can interpret data on transpiration rates.	
I can explain why plants need nitrates, phosphates, potassium and magnesium.	
I can relate mineral deficiencies to their symptoms.	
I can describe the effects of temperature, oxygen and water on the rate of decay.	
I can explain how food preservation methods reduce the rate of decay.	
I can describe how plants can be grown without soil (hydroponics).	
I can explain the advantages and disadvantages of biological control.	
I am working at grades D–C	

I can explain what it means for an ecosystem to be self supporting.	
I can describe zonation of species across a habitat.	
I can explain how isotopes have increased our understanding of photosynthesis.	
I can explain the effects of limiting factors on the rate of photosynthesis.	
I can explain how the cellular leaf structure is adapted for efficient photosynthesis.	
I can interpret data on the absorption of light by photosynthetic pigments.	
I can explain how the rate of diffusion can be increased.	
I can predict the direction of water movement in osmosis.	
I can describe the structure of xylem and phloem.	
I can explain how the cellular leaf structure is adapted to reduce water loss.	
I can describe how mineral elements are used to produce useful plant compounds.	
I can explain how root hairs are taken up by root hairs using active transport.	
I can explain why changing temperature, oxygen and water affect the rate of decay.	
I can explain how saprophytic fungi use extracellular digestion.	
I can explain the advantages and disadvantages of hydroponics.	
I can explain how intensive food production improves the efficiency of energy transfer.	
I am working at grades B–A*	

B5 Grade booster checklist

B5	
I can describe the different types of bone fractures.	
I can identify the main bones and muscles in a human arm.	
I can understand the differences between open and closed circulatory blood systems.	
I can describe the pulse as a measure of the heart beat to put blood under pressure.	
I can recall that there are different blood groups called A, B, AB and O.	
I can describe reasons for blood donation.	
I can understand why most living things need oxygen to release energy from food.	
I can describe the functions of the main parts of the human respiratory system.	
I can describe the position and function of the main parts of the human digestive system.	
I can understand that in chemical digestion, enzymes break down large food molecules.	
I can explain the difference between egestion and excretion.	
I can name and locate the positions of the main organs of excretion (lungs, kidney, skin).	
I can describe the main stages of the menstrual cycle.	
I can understand reasons for checking foetal development.	
I can recall that growth can be measured as an increase in height or mass.	
I can describe the main stages of human growth and identify them on a growth curve.	
I am working at grades D–C	

I can explain the advantages of an internal skeleton compared to an external skeleton.	
I can describe how the biceps and triceps muscles operate as antagonistic muscles.	
I can understand how heart muscle contraction is controlled by pacemaker cells.	
I can interpret data on pressure changes in arteries, veins and capillaries.	
I can explain the consequences of a 'hole in the heart'.	
I can describe the processes of blood donation and blood transfusion.	
I can explain the terms tidal air, vital capacity air, residual air and total lung capacity.	
I can describe the symptoms of asthma and its treatment.	
I can explain how carbohydrates, proteins and fats are digested by specific enzymes.	
I can describe how small digested food particles are absorbed.	
I can describe the gross structure of a kidney and associated blood vessels.	
I can explain why carbon dioxide must be removed from the body.	
I can describe the role of hormones in the menstrual cycle.	
I can explain the arguments for and against specific infertility treatments.	
I can explain possible causes of the increase in life expectancy in recent times.	
I can explain problems in the supply of donor organs.	
I am working at grades B–A*	

B6 Grade booster checklist

B6	
I can explain how the parts of bacterial cells are related to their function.	
I can describe the factors that affect the growth rate of yeast.	
I can describe how the transmission of diseases can be prevented.	
I can describe how antibiotics and antiseptics can be used in the control of disease.	
I can describe the main stages in yogurt making.	
I can describe the main stages in brewing beer or wine.	
I can describe the advantages of using biofuels rather than fossil fuels.	
I can describe how biogas and gasohol are produced.	
I can describe experiments to analyse the composition of soil samples.	
I can explain why earthworms are important in maintaining soil structure and fertility.	
I can explain the advantages of living in water.	
I can explain the causes of eutrophication.	
I can describe the range of enzymes that are used in biological washing powders.	
I can describe how enzymes can be immobilised.	
I can describe the main stages in genetic engineering.	
I can interpret data on DNA 'fingerprinting'.	
I am working at grades D–C	

I can explain why bacteria are so successful.	
I can explain how viruses reproduce.	
I can interpret data on the incidence of certain diseases.	
I can explain precautions that can be taken to prevent the spread of antibiotic resistance.	
I can describe the action of *Lactobacillus* in yogurt production.	
I can explain the effect of various conditions on the rate of fermentation of yeast.	
I can explain some of the problems that have been produced by the cultivation of biofuels.	
I can explain some of the advantages and disadvantages of using biogas.	
I can explain the results of soil experiments on air content and permeability.	
I can explain why aeration and draining will improve soils.	
I can explain the problems of water balance caused by osmosis.	
I can explain the variety of producers in marine food chains.	
I can explain how biological washing powders work.	
I can explain the symptoms of lactose intolerance.	
I can describe the action of enzymes in genetic engineering.	
I can describe how to produce a DNA fingerprint.	
I am working at grades B–A*	

Index

B1 Answers

B1 Understanding organisms

Page 74 Fitness and health

1 a Take more regular exercise; eat a healthier diet; reduce salt intake; drink less alcohol; avoid stress; don't smoke; maintain a healthy weight *(Any 2)*

b With high blood pressure, increased risk of small blood vessels bursting; if in brain a stroke happens; if in kidney, kidney damage occurs

2 a Risk factors of hypertension, smoking and high cholesterol show overall decrease; risk factor of being overweight has increased

b Three out of four risk factors reduced, so expect reduction in incidence of heart disease; however, being overweight shows a large increase, so overall effect may mean little change; change in risk factors will take many years to show effect *(Any 2)*

3 Being fit is the ability to do exercise – this will not stop/kill pathogens

4 Carbon monoxide reduces the oxygen capacity of red blood cells; it combines with the haemoglobin in red blood cells; preventing combination with oxygen/ oxy-haemoglobin; so the blood contains less oxygen; can be fatal *(Any 4)*

Page 75 Human health and diet

1 a Kwashiorkor

b Overpopulation; limited investment in agriculture

2 a 40000 g = 40 kg; EAR = 40 × 0.6 = 24 g

b May be vegetarian or vegan and not ensuring adequate nutrition; may have poor self-image

c Plant proteins are second-class proteins, i.e. *one* plant source alone does not contain all the essential amino acids; the body cannot make its own essential amino acids; so Simon will need to eat a combination of plant foods that make a first-class protein, e.g. beans on toast *(Any 2)*

3 a Carbohydrates

b Amino acids

c Extra glycogen is stored in liver; excess protein cannot be stored

Page 76 Staying healthy

1 a Pathogens; toxins

b Antigens; antibodies

2 a Parasite: mosquito/*Plasmodium;* host: human

b Drain stagnant water to kill larvae of mosquito/vector; put oil on water to prevent larvae of mosquito/vector breathing; spray insecticide to kill adult mosquito/ vector; take medication to kill *Plasmodium* inside body *(Any 2)*

3 a i At 20 days the level of immunity from passive immunity is twice that of/much higher than active immunity

ii After 60 days there is no immunity from passive immunity, active immunity level is still high

b In passive immunity the body receives antibodies; this results in a high immunity level being quickly reached; in active immunity the body makes its own antibodies continually; so the level of antibodies and immunity remains high

c Passive immunity (no mark) because immunity response is quicker

d Antibiotics are used to kill bacteria; bacteria show differences, some will be resistant to the antibiotic; these resistant strains will survive due to less competition; so new antibiotics will need to be developed since existing antibiotics will not work *(Any 3)*

Page 77 The nervous system

1 a Reflected should be refracted; optic nerve should be retina; spinal cord should be brain

b Narrower field of view; poorer judgement of distance

c Explanation of accommodation, i.e. ability to change focus; for distant objects the lens is made thinner; by the suspensory ligaments becoming tighter; and the ciliary muscles relaxing; accept explanation for near objects

2 a A: cell body; B: axon; C: sheath

b Axon

c C

d Nerve impulse triggers release of neurotransmitter substance; which diffuses; across synapse; and binds with receptor molecules; in membrane of next neurone; triggering new impulse *(Any 5)*

Page 78 Drugs and you

1 Depressant linked to alcohol; painkiller linked to paracetamol; stimulant linked to caffeine; hallucinogen linked to LSD

2 a i 251

ii Fewer men smoke; people now know risks/publicity/ warnings on cigarette packets/ban in public areas

b Cause more neurotransmitter substances; to diffuse across synapse; more nerve impulses transmitted

3 a Reading, texting and having alcohol all increase reaction distances; at both 35 and 70 mph; greater effect at 70 mph; greatest effect from reading, then texting, then alcohol; using alcohol roughly doubles reaction distances *(Any 3)*

b Not reliable since only one person tested; condition of car/tyres/road not known

Page 79 Staying in balance

1 a Maintaining a constant internal environment

b Optimum temperature for action of many enzymes

c Sweat more; therefore lose too much water

d i A negative feedback mechanism cancels out a change; it returns it to normal

ii Give an example, e.g. body too hot so blood temperature rises; hypothalamus in brain detects this change; triggers sweating and vasodilation

2 a Type 2 diabetes caused by too little insulin or by cells not reacting to it; amounts of carbohydrates eaten can be altered; to suit activity

b i Glucose enters blood

ii Insulin; converts excess glucose into glycogen

iii No insulin; so glucose remains in blood

Page 80 Controlling plant growth

1 a Makes roots grow; because rooting powder contains auxins/plant hormones

b i Because they kill only some specific weeds

ii Contains plant hormones; which makes plant/roots grow too fast

2 a Positively; gravity; geotropic; auxins

b i Auxins

ii Tip

iii Shoot curves/grows away; from the side which has the agar block; agar block contains auxin; which diffuses into the left side of the shoot; higher concentration of auxin on left side; causes more cell elongation on left side *(Any 5)*

Page 81 Variation and inheritance

1 a Sudden change in gene/chromosome

b In gamete formation; genes mixed up; in fertilisation; recombination of genes from two parents *(Any 2)*

2 23; a different; the same

3 a Correct lengths of chromosomes; correct letters; correct gender

X X female　X Y male　X X female　X Y male

b Two sex chromosomes X and Y; these separate in gamete formation; male is XY, female is XX; ratio from cross is 1:1 *(Any 3)*

c i Correct gametes; correct combinations (as shown in diagram)

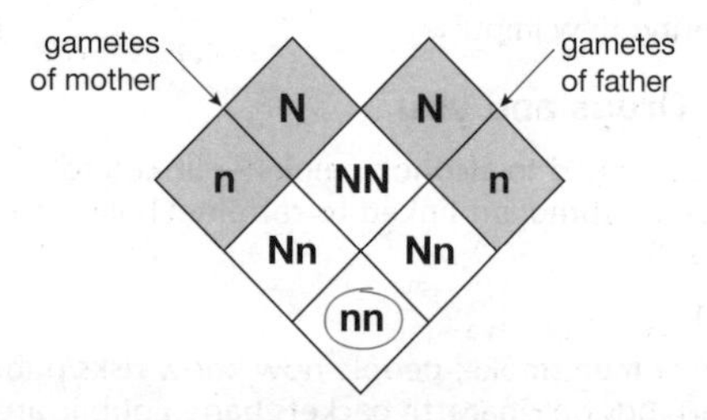

ii nn ringed (as shown in diagram)

Page 82 B1 Extended response question

5–6 marks

Uses chart to work out that Rick has an ideal blood pressure, Shabeena is pre-high blood pressure while Alan and Toni have high blood pressure. Applies knowledge of how a poor life style (being overweight, stress, high alcohol intake, smoking) increases blood pressure and a healthy lifestyle (regular exercise, balanced diet) lowers blood pressure to ideal levels. Realises that Alan, Shabeena and Toni have high risk of stroke/kidney damage. Links other information on the risks of smoking (lung cancer) and poor diet (diabetes) to the data. All information in answer is relevant, clear, organised and presented in a structured and coherent format. Specialist terms are used appropriately. There are few, if any, errors in grammar, punctuation and spelling.

3–4 marks

Correctly applies data to blood pressure chart. Gives a limited description of some life style factors and consequence but links to actual people not clear. Links to smoking risks and poor diet brief or missing. For the most part the information is relevant and presented in a structured and coherent format. Specialist terms are used for the most part appropriately. There are occasional errors in grammar, punctuation and spelling.

1–2 marks

Errors in applying data to the blood pressure chart. An incomplete/confused description is given of reasons and consequences for data. Answer may be simplistic. There may be limited use of specialist terms. Errors of grammar, punctuation and spelling prevent communication of the science.

0 marks

Insufficient or irrelevant science. Answer not worthy of credit.

B2 Understanding our environment

Page 83 Classification

1 a Phylum; class; order; genus; species

b DNA analysis; to see how similar it is

2 a A group of organisms that can interbreed; to produce fertile offspring

b i Bobcat and ocelot

ii Both belong to the same genus

3 It has some features of a reptile and some of a bird; an organism that has features of more than group (like this one, which has feathers and also has teeth) will be difficult to classify

4 a Hybrid

b It is infertile/has some of the characteristics of a horse and some of a zebra

5 They share a fairly recent common ancestor; therefore have similar genes; both live in similar habitats; have developed similar adaptations *(Any 3)*

Page 84 Energy flow

1 a i the dry mass of living material; at each trophic level

ii one rose bush can feed many beetles; pyramid of numbers does not take into account the size of the organism

b Measuring biomass involves removing all the water; destructive process; need to collect all the parts of the rose bush *(Any 2)*

2 a Respiration

b i Keep the cows in warm conditions; restrict their movement

ii have more energy available for growth

c i 3056 – (1909 + 1022); = 125kJ

ii $\frac{125}{3056} \times 100$; = 4.1%

iii So much energy is lost between each trophic level; if we ate crops there would only be one energy transfer; less energy would be lost *(Any 2)*

Page 85 Recycling

1 a Combustion; respiration; photosynthesis

b Less oxygen in waterlogged soils; decomposers cannot respire

c Turn into limestone; limestone weathered by acid rain; chemical reaction releases carbon dioxide

2 a To make proteins

b i Ammonia　**ii** Nitrifying bacteria

c Nitrogen-fixing bacteria; in soils; or in root nodules of legumes; convert nitrogen to ammonia or nitrates; action of lightning; combines oxygen and nitrogen *(Any 4)*

Page 86 Interdependence

1 a Mates

b i Where it lives; its role in the community

ii They both live in the same part of the forest; both eat acorns

2 a The numbers go up when there are fewer owls because fewer lemmings are eaten; when the owl numbers increase the lemming numbers drop as more are eaten

b When there are lots of lemmings, then owls can survive and breed; it takes a while for owl numbers to increase

3 Mutualism

4 a Mites

b Bees gain nectar; flowers get pollinated

c Pea plant gives the bacteria some sugars made by photosynthesis; the bacteria fix nitrogen into compounds that the plant can use

Page 87 Adaptations

1 a Thick fur for insulation; layer of blubber under the skin; wide feet stop it sinking into the snow; claws to hold on to the ice; white colour for camouflage; small ears lose less heat; large body to retain heat *(Any 3)*

b Hibernation; body reactions slow down; conserve food reserves; less food is available in winter *(Any 3)*

- c Polar bears live in colder climates; smaller ears/larger body means smaller surface area to volume ratio; so less heat loss

2 a Less water loss through leaves; protect cacti from being eaten
- b Sun basking/absorb Sun's heat; to warm body up

3 If one food is in short supply they can change to another; different foods are available in different areas

Page 88 Natural selection

1 a Can digest the acorns more easily
- b In the genes; which are passed on in the sex cells
- c Some of the red squirrels would have a mutation; this would allow them to digest acorns; they are more likely to survive; pass on this gene; over many generations the population will all be able to digest *(Any 4)*

2 Some bacteria are resistant to antibiotics; they can survive and reproduce

3 a Worried that people would disagree with him; most people believed that God created all organisms as they are now
- b Organisms on small islands do not grow as fast; pass this on to their offspring
- c Acquired characteristics are not inherited

Page 89 Population and pollution

1 a Increase in temperatures/global warming/greenhouse effect; rising sea levels/flooding; drought *(Any 2)*
- b i CFCs
  - ii UV light causes skin cancer

2 a Exponential
- b Burn less fossil fuels; produce less waste that cannot be recycled; eat less/local food; less travel *(Any 2)*

3 a Indicator species
- b Mussels, damselfly larvae and bloodworms all survive in polluted water; if it was clean you would also find mayfly and stonefly larvae
- c Advantage: more accurate result at any one time; Disadvantage: more expensive; does not monitor over a very long period

Page 90 Sustainability

1 a More tourism; better transport; other resources
- b To see how closely related they are; so prevent inbreeding/maintain genetic variation

2 a For: provides food/jobs; Against: may lead to their extinction
- b Areas too large to police

3 a Taking enough resource from the environment to supply an increasing population but leaving enough behind to ensure a supply for the future and prevent permanent damage
- b Restricting numbers stops too many being killed; preventing killing small fish allows them to mature to breed
- c It is difficult to dispose of the increased waste produced without causing pollution; the people need more food which needs to be provided without destroying habitats; increased demand for energy which could cause pollution if generated from fossil fuels

Page 91 B2 Extended response question

5–6 marks

A description of how energy is lost from a food chain and some realisation that efficiency of transfer is different between different trophic levels. Includes an appreciation that vegetarianism involves fewer transfers of energy and so less energy loss. Answers are backed up by at least one calculation of efficiency. All information in answer is relevant, clear, organised and presented in a structured and coherent format. Specialist terms are used appropriately. Few, if any, errors in grammar, punctuation and spelling.

3–4 marks

Descriptions of how energy is lost from food chains and an appreciation that vegetarianism would provide humans with more energy. Figures are quoted but not processed. For the most part the information is relevant and presented in a structured and coherent format. Specialist terms are used for the most part appropriately. There are occasional errors in grammar, punctuation and spelling.

1–2 marks

An incomplete description, naming some processes by which energy is lost but no analysis of figures or appreciation of the importance of the number of trophic levels. Answer may be simplistic. There may be limited use of specialist terms. Errors of grammar, punctuation and spelling prevent communication of the science.

0 marks

Insufficient or irrelevant science. Answer not worthy of credit.

B3 Living and growing

Page 92 Molecules of life

1 a To provide lots of energy; for contraction
- b In the cytoplasm
- c They are the site of protein synthesis

2 a i A circle around any one of the squares on the diagram
  - ii Double helix
  - iii Proteins are made in the cytoplasm; DNA cannot leave the nucleus
- b i 5 ii ATATACATTTTGTT

3 a C always equals G; T always equals A
- b They realised that C always bonds with G and T with A; this holds the two chains together

Page 93 Proteins and mutations

1 a Collagen – a structural protein, haemoglobin – a carrier protein, insulin – a hormone
- b Amino acids
- c They have a different order of amino acids; so the chains fold up differently *(Either)*

2 a Enzymes are specific; it would be the wrong shape to fit the active site
- b i As the temperature increases the reaction is faster; after a certain temperature any increase will decrease the rate
  - ii 41–42 °C iii $Q_{10} = \frac{4}{2} = 2$
  - iv Increased movement of the molecules; which leads to more collisions

3 a Radiation; certain chemicals
- b Sometimes they produce an advantage; this is more likely to be passed on
- c Could be a change in the gene coding for enzyme B; a different base sequence might lead to a different order of amino acids; may stop enzyme B working so the red pigment is not converted to purple

Page 94 Respiration

1 a ATP traps the energy released by respiration; it passes it on to other processes that need it
- b $6O_2 \rightarrow 6CO_2 + 6H_2O$
- c i As the horse runs faster there is an increase in lactic acid levels; this is slow at first but increases more rapidly after about 8 m/s

ii At slower speeds respiration is mainly aerobic; at high speeds there is more anaerobic respiration and so more lactic acid is made

iii About 9 m/s

d Gets sent to the liver; broken down with the use of oxygen

2 a Measure the rate of oxygen consumption/rate of carbon dioxide production

b Respiration is controlled by enzymes; high temperatures will stop them working

Page 95 Cell division

1 a Any two from: allows organism to be larger; allows for cell differentiation; allows organism to be more complex

b i 2.0; 1.5; 1.2

ii They have a smaller surface area to volume ratio; this reduces the rate of diffusion

2 a Before cells divide, DNA replication takes place; the new cells are diploid

b The two original strands come apart; each one acts as a template attracting complementary bases

3 a The acrosome is needed to digest the egg membrane; this allows the nucleus of the sperm to enter the egg

b In the first stage there are two pairs of chromosomes; each member of the pair moves apart; in the second division there is only one chromosome from each pair and the copies move apart

Page 96 The circulatory system

1 a Disc shaped: larger surface area to take up oxygen quicker; no nucleus: more room to carry more haemoglobin/oxygen

b Haemoglobin

c It combines with oxygen in the lungs forming oxyhaemoglobin; this is reversible; oxygen is released from oxyhaemoglobin in the tissues

2 a Arteries carry blood away from the heart; veins carry blood back to the heart; capillaries are the site of exchange from the blood

b i A = artery; B = vein; C = capillary

ii Wall is only one cell thick; this makes it permeable

3 a

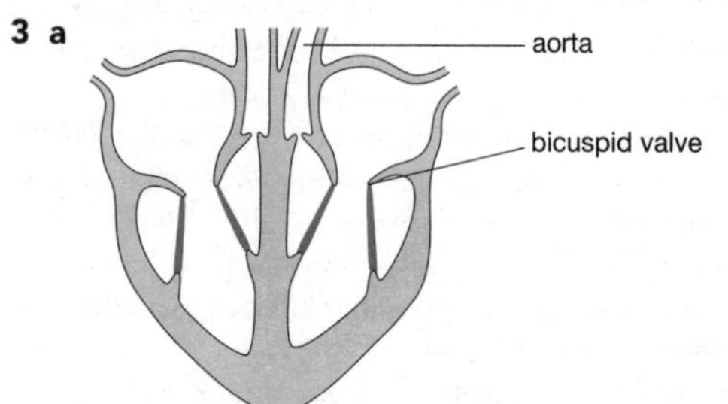

b The left ventricle has to pump the blood further; it has to generate more pressure

c Increased pressure so blood flows faster/more oxygen to the tissues

Page 97 Growth and development

1 a A bacterial cell does not have a true nucleus; or mitochondria

b It is circular rather than a strand

2 a Adolescence/puberty

b The girl is taller; because girls reach puberty/have their growth spurt younger

c Advantages: quick/not destructive; disadvantage: only measures growth in one dimension

3 a Stem cells

b Embryonic stem cells can form any type of cell; adult stem cells are more restricted

4 The parts of a plant where cells divide

Page 98 New genes for old

1 a When two individuals that are closely related mate

b Lack of genetic variation; leads to health problems

2 a Two from: many people are deficient in Vitamin A; this can be made from beta-carotene; many of these people eat large amounts of rice

b The plants might be harmful to health in the long term; they might escape into the wild and affect food chains

c The gene for beta-carotene would have to be identified; then cut out of the carrot DNA; inserted into the DNA of rice

3 a Gene therapy

b For: may cure life-threatening diseases; against: may be dangerous for the patient; some people think that it is ethically wrong to move genes between organisms

Page 99 Cloning

1 a Putting the nucleus from a body cell into an egg that has had its nucleus removed

b i The pigs have human genes; so their organs are not rejected

ii Produce large quantities of pigs with desired characteristics such as meat yield/produce large numbers of pigs which produce human proteins

2 a Given an electric shock

b Julie; because it contains her DNA

3 a Advantages: produce large numbers quickly/if the parent produces good strawberries then the offspring will also do so; disadvantage: all the strawberries will be identical and so could all be attacked by a disease

b Aseptic technique means making sure that no microbes infect the plants; the growth medium is a gel that contains all the nutrients that are needed for the plant to grow

Page 100 B3 Extended response question

5–6 marks

The answer includes an explanation of how a mutation can change a base sequence, hence the amino acid sequence and so collagen structure. It also includes an explanation of the symptoms in terms of collagen's role as a structural protein. The particular susceptibility of arteries due to the high pressure is mentioned. All information in answer is relevant, clear, organised and presented in a structured and coherent format. Specialist terms are used appropriately. Few, if any, errors in grammar, punctuation and spelling.

3–4 marks

The answer refers to mutations and changes in DNA or genes but the precise mechanism is not clear. The function of collagen as a structural protein is appreciated. For the most part the information is relevant and presented in a structured and coherent format. Specialist terms are used for the most part appropriately. There are occasional errors in grammar, punctuation and spelling.

1–2 marks

An incomplete description, either showing an appreciation of mutations and their inherited nature or an understanding of the role of collagen. Answer may be simplistic. There may be limited use of specialist terms. Errors of grammar, punctuation and spelling prevent communication of the science.

0 marks

Insufficient or irrelevant science. Answer not worthy of credit.

B4 It's a green world

Page 101 Ecology in the local environment

1 a S. balanoides

b Zonation

- **c** Exposure/temperature; causing drying out, some species more resistant; OR physical action of waves; some species attached more strongly

2 a Natural ecosystems have bigger diversity; native woodlands/lakes are natural ecosystems; forestry plantations/fish farms are artificial ecosystems

- **b** Plants depend on animals for pollination/fertilisation; animals depend on plants for food/idea of food chains; plants produce oxygen, animals use it for respiration; animals produce carbon dioxide, plants use it for photosynthesis; idea of recycling of minerals after death

3 a Capture–recapture

- **b i** $\frac{50 \times 60}{10}$; population size is 300
- **ii** Not all ladybirds are counted/population is sampled
- **c** No deaths, immigration, emigration; marking does not affect survival; identical technique used *(Any 4)*

Page 102 Photosynthesis

1 a i $C_6H_{12}O_6$; $6O_2$ **ii** Light/Sun

- **b i** Cell walls; starch; growth/repair; storage
- **ii** Energy source/respiration

2 a Plants produce oxygen

- **b** Isotope of oxygen $^{18}O_2$ used as part of oxygen molecule; in photosynthesis the isotope given off as oxygen gas; so water was split up by light energy not carbon dioxide

3 Heater produces warmth; and carbon dioxide; shades removed so maximum light enters; watering system provides right amount of water

4 a Carbon dioxide levels drop at midday, oxygen levels rise at midday; because carbon dioxide is used up and oxygen is released in photosynthesis; the use of oxygen and production of carbon dioxide by goldfish is constant; so it does not affect the shape of the graphs *(Any three)*

- **b** Levels of carbon dioxide would increase; levels of oxygen would decrease; due to goldfish respiration and no photosynthesis

Page 103 Leaves and photosynthesis

1 a Shorter distance for gases to diffuse/all cells can get light for photosynthesis; variety of pigments; vascular bundles/xylem and phloem; open and close stomata

- **b i** Spongy mesophyll; palisade
- **ii** Eats cells containing chlorophyll; therefore parts will not be green

2 a chlorophyll a and b

- **b** Parts of light spectrum used (in photosynthesis)
- **c** Different pigments/more than one pigment; each absorbs different parts of the light spectrum
- **d** Green/about 550 nm
- **e** 400–500 nm approx; 640–670 nm approx
- **f** Little light will reach them; only blue light will reach them; little photosynthesis/poor or no growth or need very large leaves

Page 104 Diffusion and osmosis

1 a Black circles move down and mix with white circles; white circles move up and mix with black circles

- **b** Diffusion
- **c** Random movement of particles; takes place from high to low concentrations
- **d** A shorter distance to travel/nose nearer bottle; a higher concentration gradient/stronger perfume; a greater surface area/wider neck of bottle

2 Water; partially permeable; dilute; concentrated; random

3 a Water leaves cells; contents/cytoplasm shrinks/pulls away from cell wall

- **b** Cells collapse; plant wilts
- **c** Animal cells do not have cell walls; animal cells become creanate

Page 105 Transport in plants

1 a Vascular; xylem; phloem

- **b** Xylem cells are dead (phloem cells are living); have extra cell wall thickening; have a hollow lumen

2 a 2, 9

- **b** Leaf A (no mark), fewer stomata open; no photosynthesis/stomata close to prevent too much water loss
- **c** Water enters guard cells; turgidity increases; guard cells curve

3 a Water loss/transpiration

- **b** Leaf 4 has only lower surface exposed; it loses more weight/2.7 g instead of 1.0 g; it loses more water than Leaf 3, which has upper surface only exposed
- **c** Loses more weight; because loses more water; since rate of evaporation of water increased

Page 106 Plants need minerals

1 a Nitrates; phosphates; potassium; magnesium

- **b** Cannot make proteins; for cell growth

2 Make amino acids/proteins; make DNA/cell membranes; make chlorophyll

3 a Active transport

- **b** Uses energy; minerals moved against a concentration gradient; minerals are selected; uses a carrier system

Page 107 Decay

1 a A temperature of 25 °C; plenty of oxygen

- **b** Earthworms; maggots; woodlice
- **c** For: avoids landfill/recycles minerals/saves buying fertilisers; Against: unsightly/smelly if not working properly (anaerobic respiration)/needs space or garden
- **d i** Break up dead material; so increasing surface area for decay
- **ii** Digest dead material; by extracellular digestion/digesting food outside the body and absorbing it
- **e** Increase in temperature increases growth and reproduction; of microbes/bacteria; by increasing enzyme action in respiration; extremes of temperature will slow down microbial growth; because enzymes will not work/be denatured

2 a Temperature of 5 °C slows down microbial growth/reproduction; so slowing down decay

- **b** Temperature of –22 °C kills most bacteria/microbes; so slowing down decay better than a refrigerator

Page 108 Farming

1 a Hydroponics

- **b i** Better control of mineral levels; better control of diseases; can be used in areas of poor/barren soil *(Any two)*
- **ii** Lack of support for tall plants; still uses fertilisers; needs electricity supply *(Any two)*

2 a Biological control

- **b** Insecticides can enter and accumulate in food chains; insecticides may harm other useful insects; some insecticides are persistent *(Any two)*
- **c** Not native/from South America; no natural predators; lack of research on its diet/lack of trials *(Any two)*

3 a To farm organically; to avoid build-up of pests

- **b** Longer crop time/succession of crops; avoid certain times when insect pests hatch out/most abundant

Page 109 B4 Extended response question

5–6 marks

The answer includes an explanation of osmosis and the importance of the partially-permeable membrane. It correctly predicts the movement of water molecules towards the left, i.e. inside the cell. It explains the importance of the size of molecules, the large sugar molecule remaining inside the cell. The answer includes a reference to the importance of the process, i.e. uptake of water into the cell, maintaining turgor, providing water for photosynthesis as well as the cell being able to retain food molecules such as sugar. All information in answer is relevant, clear, organised and presented in a structured and coherent format. Specialist terms are used appropriately. Few, if any, errors in grammar, punctuation and spelling.

3–4 marks

The answer refers to osmosis and makes some attempt to predict movement of molecules. References to partially-permeable membrane or sugar retention may be missing. For the most part the information is relevant and presented in a structured and coherent format. Specialist terms are used for the most part appropriately. There are occasional errors in grammar, punctuation and spelling.

1–2 marks

An incomplete description of osmosis is given and only a vague reference to movement of molecules. The importance of water movement is briefly mentioned without details. Answer may be simplistic. There may be limited use of specialist terms. Errors of grammar, punctuation and spelling prevent communication of the science.

0 marks

Insufficient or irrelevant science. Answer not worthy of credit.

B5 The living body

Page 110 Skeletons

1 a Contain living cells; are hollow; contain bone marrow

b Cartilage replaced by bone; by adding calcium/phosphorus; some cartilage left on bone heads; some cartilage left between heads and shaft when still growing

2 a i Ball and socket **ii** Hinge

b Acts as cushion/absorbs shock; provides lubrication

c Biceps and triceps are antagonistic muscles; as biceps contracts, triceps relaxes; producing precise control; to raise forearm: or reverse argument to lower forearm

d Multiplies effort of muscle contraction; elbow acts as fulcrum

3 Lack of calcium and phosphorus in bones of elderly people; so long bones not as strong/low density; higher risk of fracture

Page 111 Circulatory systems and the cardiac cycle

1 a Single/closed; two; double; four

b Creates a higher blood pressure; enabling faster distribution of food and oxygen; separates deoxygenated and oxygenated blood

2 a As heart muscle in ventricles contracts, higher blood pressure; as heart muscle relaxes , lower blood pressure

b High in arteries because of muscle contraction in left ventricle; blood travels around the body before it reaches veins

c Walls are very thin/walls only 1 cell thick; high pressure would burst walls; allows for gas exchange

3 a Second century: knew heart acted as a pump; knew pulse was a indication of heart beat; but thought liver made blood; blood flowed backwards and forwards *(Any two)*

Seventeenth century: blood circulated around body; heart of four chambers; knew there were tiny blood vessels connecting arteries and veins *(Any two)*

b Better microscopes; better technology such as body scanning/ECG

4 P shows impulses from SAN; R shows impulses from ventricles (AVN); T shows impulses as ventricles contract

Page 112 Running repairs

1 a Blood can move directly from right to left ventricle; so deoxygenated and oxygenated blood will mix; so blood leaving the heart will contain less oxygen than normal; blood pressure will be lower due to blood flowing back into right ventricle *(Any three)*

b Blood contains less oxygen; so blood not as bright red

c Lungs in unborn baby not working; since baby surrounded by fluid; so blood is oxygenated at placenta not lungs; so heart structure/blood circulation changes at birth *(Any three)*

2 a Donor; transfusion; heparin/aspirin/warfarin

b Blood platelets in contact with air/damaged blood vessels; series of chemical reactions/cascade process; mesh of fibrin fibres form

3 a Anti B; anti A; A and B; none

b Antigens; antibodies

c Blood clumping; when incompatible blood is mixed; example given blood group A given to blood group B

Page 113 Respiratory systems

1 a Life style choice to lung cancer; genetic to cystic fibrosis; exposure to fibres to asbestosis
(one correct = one mark; two or three correct = two marks)
Asbestosis to inflammation/scarring of lungs; cystic fibrosis to too much mucus; lung cancer to rapid cell growth
(one correct = one mark; two or three correct = two marks)

b Sticky mucus traps foreign particles; cilia beat and move mucus away

c Difficulty in breathing; wheezing; tight chest

d Lining of airways become inflamed; fluid builds up in airways; muscles around bronchioles contract

2 Larger tidal air exchange/more air breathed in and out; faster rate of breathing; since more oxygen needed for respiration

3 a Volume decreases; pressure increases

b Balloon smaller/deflated

c Bell jar wall is fixed/shape of rib cage changes due to intercostal muscles

Page 114 Digestion

1 Breaks down food providing a larger surface area; for enzyme action; allows food to pass through digestive system more easily

2 a Meat broken down/smaller pieces; chemically digested; enzyme/protease digests protein; into smaller molecules/amino acids; which are soluble and dissolve; broken down by stomach acid *(Any four)*

b Stomach is acid/low pH; mouth and small intestine alkaline/high pH; because different enzymes present; and have different optimum pH values *(Any three)*

3 a Glucose; amino acids

b Fatty acids

c Fatty acids are not soluble in water/plasma; therefore would block up blood vessels; glucose and amino acids are soluble

4 Fats not emulsified; to provide a large surface area; so fat digestion will be slow/incomplete

5 Starch a complex molecule of many units/polysaccharide; broken down to units of two molecules/disaccharides/ maltose; then broken down into one unit/glucose/ monosaccharide; by different enzymes

Page 115 Waste disposal

1 a A cortex; B medulla; C renal artery; D renal vein; E ureter
   b Filter unit of glomerulus and capsule; region of selective reabsorbtion; region of salt and water regulation
2 a Less protein in urine; less glucose in urine; more urea in urine; more salt in urine; more water in urine
   b Protein and some salt reabsorbed; as useful to the body; urea, excess salt and water not reabsorbed; so they can be excreted
   c Amount of water/salt reduced; because water lost in sweating
   d water uptake; exercise
3 a Toxic at high levels
   b Detected by detectors in carotid artery/brain; breathing rate/heart rate increased
4 Negative feedback corrects a change; concentration controlled by hormone ADH; ADH stops being secreted when concentration of blood decreases; so kidney tubules reabsorb less water; and dilute urine is excreted *(Any four)*

Page 116 Life goes on

1 a Oestrogen to repair of uterus wall; progesterone to maintain uterus wall; FSH to stimulate egg to develop; LH to control ovulation *(one correct = one mark; two correct = two marks; three or four correct = three marks)*
   b After fertilisation, levels of oestrogen do not increase; since uterus wall will not be broken down; levels of progesterone will remain high; to maintain the uterus wall
   c When oestrogen levels are high, negative feedback to reduce FSH levels; this results in the ovary follicle not being stimulated; so oestrogen level falls
2 IVF: multiple births/lower birth weight; Egg donation: not genetic code of mother; Artificial insemination: sperm and egg not put in direct contact; Surrogacy: become emotionally attached to baby; Use of FSH: injections/ treatments; Ovary transplants: shortage of donors/rejection

Page 117 Growth and repair

1 a Shortage of donors; tissue match; size/age of donor organ
   b Acceptance to use part of a dead body; waiting/relying on the death of someone; should it have been done just to find out if it was possible; justification of use of resources for one person *(Any two)*
   c Rejection; having to use immuno-suppressive drugs for lifetime; psychological problems, e.g. limb not belonging to recipient *(Any two)*
   d More chance of finding a suitable donor worldwide; immediately known if donation is available
2 a Kidney
   b Yes: high number donated and grafted/No: no information on success rate
   c Twice the number of transplants than donors; dead person can donate two kidneys
   d Major operation/high risk/rejection results in death
3 Less industrial disease; healthier diet; better lifestyle; modern treatments; modern cures for disease; better housing *(Any three)*

Page 118 B5 Extended response question

5–6 marks
The answer includes correctly labelled diagrams to show the two different systems. The systems are correctly identified (single/double) and described. It correctly links a two-chambered heart to a single circulatory system and a four-chambered heart to a double circulatory system and clearly explains the link. The consequent higher blood pressure (and more powerful heart muscle) in humans is realised together with the improved food and oxygen distribution. Detail on human heart structure clear and accurate. A separate human heart diagram may be drawn. May mention different circulation in unborn babies. All information in answer is relevant, clear, organised and presented in a structured and coherent format. Specialist terms are used appropriately. Few, if any, errors in grammar, punctuation and spelling.

3–4 marks
The diagrams are correct, though they may not be fully labelled. Links between the number of heart chambers and type of circulatory system are made but explanation is vague or lacking. Little information on human heart structure and function. For the most part the information is relevant and presented in a structured and coherent format. Specialist terms are used for the most part appropriately. There are occasional errors in grammar, punctuation and spelling.

1–2 marks
Diagrams very poor and lacking in headings/labels. Some realisation of different heart structure but explanations lacking. Answer may be simplistic. There may be limited use of specialist terms. Errors of grammar, punctuation and spelling prevent communication of the science.

0 marks
Insufficient or irrelevant science. Answer not worthy of credit.

B6 Beyond the microscope

Page 119 Understanding microbes

1 a Rod; spiral; curved rods *[Any two]*
   b Binary fission
2 a Makes us ill before the immune system can destroy them
   b Prevent infection
   c To prevent contamination from bacteria in the air/ prevent bacteria escaping
3 a 40 °C
   b Food availability; correct pH; no build-up of waste *[Any two]*
   c It will increase; from 38 to 70, so double the rate
4 a No cell membrane/no flagellum/no cytoplasm
   b Attaches itself to a host cell; injects its genetic material into the cell; uses the cell to make the components of new viruses; the host cell splits open to release the viruses *[Any three]*

Page 120 Harmful microorganisms

1 a Isolation/use of tissues with disinfectant
   b Incubation period
   c i The death rate equals the number of cases
      ii Not everybody who gets the disease will necessarily see a doctor/some people have natural immunity
      iii Number of deaths has decreased from 1924 to 1948; due to improved health care or hygiene; death rate decreases more rapidly after 1948; due to the introduction of antibiotics *[Any three]*
2 a Sir Alexander Fleming
   b Fungus
   c i Certain individual bacteria are not killed by antibiotics; they are more likely to survive and reproduce; the decrease in the number of cases of TB slowed after about 1960
      ii Only prescribe antibiotics when necessary; always complete the course *(Either)*

Page 121 Useful microorganisms

1 a (B) A, D, C
  b Breakdown the lactose in milk; to produce lactic acid

2 a Make the liquid clear/make the yeast settle out
  b Concentrates the alcohol; to make spirits (such as brandy)
  c Kill any harmful bacteria

3 a $C_6H_{12}O_6$; $2CO_2$
  b Both strains increase in numbers over time; A continues to increase but numbers of B peak and drop before the end; the decrease in growth may be due to the food supply running out or a build-up in waste; A may be more tolerant to high alcohol levels than B

Page 122 Biofuels

1 a No particulates are released; no increase in greenhouse gas levels
  b Can lead to destruction of habitats; and extinction of species

2 a Continuous flow
  b i The higher the temperature, the higher the volume of gas produced; up to 45 °C; when volume of gas starts to fall with increased temperature *[Any two]*
    ii As temperature increases enzymes move faster/have more kinetic energy; more collisions so increase in rate; at higher temperature enzymes start to change shape/denature; substrate will not fit into active site *[Any three]*

3 a Gasohol
  b Fermentation/anaerobic respiration
  c Higher supply of sugar; lacking in natural oil reserves

Page 123 Life in soil

1 a Humus
  b Samples of soil are dried by heating at about 100 °C; then the mass of each sample is measured; they are then strongly heated with a Bunsen burner; the samples are reweighed and the difference in mass equals the organic content *[Any three]*
  c B

2 a $\frac{2200}{3000} \times 100 = 73.3\%$
  [One mark for equation; one mark for answer]
  b Decomposition to release minerals; increase the air content
  c The worms drag leaves, etc., below the surface
  d Waterlogged soils lack air; less oxygen available for roots; so less active uptake of minerals

Page 124 Microscopic life in water

1 a Advantage = no dehydration/little temperature change/ water supports weight/easier to excrete waste; Disadvantage = difficult to maintain water balance/ difficult to move against current
  b i Water will enter their cytoplasm; by osmosis
    ii Water is pumped into a vacuole; using energy; vacuole empties water to the outside (if drawing a diagram must include above labels with same points as above) *[Any two]*

2 a Not enough light for photosynthesis
  b Bacteria → tube worms → crabs
  c Dead organic material floating down from higher levels

3 a Algae feed on the nutrients; algae population increases; algae start to die off; bacteria decompose algae using up oxygen; animals die from lack of oxygen *[Any three]*
  b Small amounts of pesticides are eaten by small organisms; larger animals feed on them/idea of a food chain; pesticides build up/are not broken down in larger animals; concentration increases enough to kill them; this is bioaccumulation *[Any three]*

Page 125 Enzymes in action

1 a
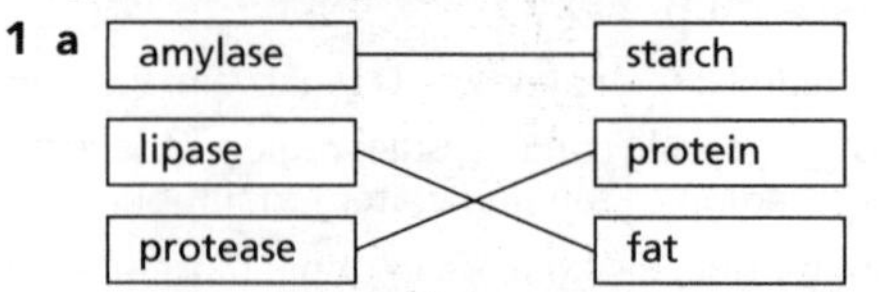

  b Products of digestion are soluble; so they get washed away

2 a Breakdown products of sucrose are sweeter than sucrose, so less is needed
  b glucose; fructose

3 a Easier to separate enzyme from product; less purification needed; enzyme can be reused; continuous process so cheaper to run *[Any two]*
  b i Lactase
    ii Cannot produce lactase; bacteria in the gut ferment lactose; this produces diarrhoea and wind

Page 126 Gene technology

1 a Transgenic
  b i Cut open the DNA of the second organism; add the DNA from the first organism to the second one
    ii Restriction enzymes
    iii A length of DNA on the end of a gene that is single stranded; it is used to join two pieces of DNA together; by complementary base pairing

2 a Gene for insulin is cut out of human DNA; gene is put into a plasmid; plasmid placed into bacteria; bacteria are cloned
  b Using an assaying technique; include gene for antibiotic resistance and then flood bacteria with antibiotic; choose the bacteria that survive *[Any two]*

3 a James; the pattern is identical
  b Extraction of DNA from sample; fragmentation of DNA using restriction enzymes; separation using electrophoresis; visualising pattern using a radioactive probe *[Any three]*

Page 127 B6 Extended response question

5–6 marks

A detailed explanation of the shape of the graph including knowledge of how the substrate can run out or waste products can limit the growth of the microbe (yeast). Shows an appreciation that adding more sugar increases the rate but does not increase the alcohol content because the alcohol kills the yeast. All information in answer is relevant, clear, organised and presented in a structured and coherent format. Specialist terms are used appropriately. Few, if any, errors in grammar, punctuation and spelling.

3–4 marks

Limited explanation of the shape of the graph including knowledge of how the substrate can run out or waste products can limit the growth of the yeast. For the most part the information is relevant and presented in a structured and coherent format. Specialist terms are used for the most part appropriately. There are occasional errors in grammar, punctuation and spelling.

1–2 marks

An incomplete explanation of the shape of the graph. Answer may be simplistic. There may be limited use of specialist terms. Errors of grammar, punctuation and spelling prevent communication of the science.

0 marks

Insufficient or irrelevant science. Answer not worthy of credit.